PLANT LIFE
IN THE WORLD'S
MEDITERRANEAN
CLIMATES

PLANT LIFE
IN THE WORLD'S
MEDITERRANEAN
CLIMATES

CALIFORNIA, CHILE, SOUTH AFRICA, AUSTRALIA,
AND THE MEDITERRANEAN BASIN

BY

PETER R. DALLMAN

Preface by Robert Ornduff

CALIFORNIA NATIVE PLANT SOCIETY

UNIVERSITY OF CALIFORNIA PRESS

Berkeley • Los Angeles

California Native Plant Society
1722 J Street, Suite 17
Sacramento, CA 95814

University of California Press
Berkeley and Los Angeles, California

Library of Congress Cataloging-in-Publication Data

Dallman, Peter R.

   Plant life in the world's mediterranean climates : California, Chile, South Africa, Australia,
and the Mediterranean Basin / Peter R. Dallman ; with a preface by Robert Ornduff.

       p. cm.

   Includes bibliographical references (p. 235) and index.

   ISBN 0-520-20808-0 (cloth : alk. paper). — ISBN 0-520-20809-9 (pbk. : alk. paper)

   1. Mediterranean-type plants   2. Plant communities.   I. Title.

QK938.M45D35 1998

581.4'2—dc21   97-52208

                                                                 CIP

Photographs by Peter R. Dallman unless otherwise noted.

Cover and book design and typesetting by Beth Hansen-Winter.

Printed in Singapore through Imago, San Juan Capistrano, California.

9 8 7 6 5 4 3 2 1

On the Cover. Center: Coastal landscape on the southern coast of Crete, Mediterranean Basin. Clockwise from top right: Tulipa doerfleri, *Mediterranean Basin;* Narcissus sp., *Mediterranean Basin;* chagual (Puya chilensis), *Chile;* pincushion protea (Leucospermum cordifolium), *Western Cape;* Douglas iris (Iris douglasiana), *California;* Eucalyptus caesia, *Western Australia;* flannel bush (Fremontodendron californicum), *California;* scarlet banksia (Banksia coccinea), *Western Australia;* pangue (Gunnera tinctoria), *Chile;* Alstroemeria sp., *Chile;* sunshine protea (Leucadendron xanthoconus), *Western Cape; and winter's bark or* canelo (Drimys winteri), *Chile.*

CONTENTS

ACKNOWLEDGMENTS   VII

PREFACE   IX

INTRODUCTION   XI

1. THE MEDITERRANEAN CLIMATE   1

   The Weather Pattern and Where to Find It
   How Location and Terrain Influence Climate
   Relationship to Other Climates
   Types of Mediterranean Climates
   Frequent Wildfires
   Weather Conditions in Relation to Plant Growth
   Unusual Weather

2. PLANT AND CLIMATE ORIGINS   19

   Floristic Kingdoms
   Early Origins of Drought-Resistant, Sclerophyll Plants
   Ice Ages and the Appearance of a Mediterranean Climate
   World Distribution of Related Plants

3. PLANT ADAPTATIONS   29

   Tough, Drought-Resistant Foliage
   Drought-Evading Foliage
   Roots with Two Water-Seeking Strategies
   Survival Following Wildfire
   Nutrient-Poor Soils
   Going Underground: Geophytes or Bulbs
   A Speedy Life Cycle: the Annuals
   Benefits of Mild, Wet Winters

4. PLANT COMMUNITIES   45

   Chaparral-like Shrublands
   Coastal Scrub
   Woodland
   Forest
   Diversity and Conservation of the Mediterranean Flora

5. CALIFORNIA 59

    Landscape and Climate
    Chaparral
    Coastal Scrub
    Oak Woodlands
    Forest

6. CENTRAL CHILE 91

    Landscape and Climate
    Matorral
    Coastal Matorral
    Woodland and Forest

7. WESTERN CAPE, SOUTH AFRICA 119

    Landscape and Climate
    Fynbos
    Strandveld
    Woodland and Forest

8. AUSTRALIA 143

    Landscape and Climate
    Kwongan
    Mallee
    Woodland and Forest

9. MEDITERRANEAN BASIN 169

    Landscape and Climate
    Maquis
    Garrigue
    Woodland and Forest

10. PLANNING A TRIP 203

    California
    Chile
    Western Cape, South Africa
    Australia
    Mediterranean Basin

REFERENCES 235

METRIC CONVERSIONS 258

ACKNOWLEDGMENTS

Robert Ornduff, Professor Emeritus of Botany and former Director of the Botanical Garden at University of California, Berkeley, has my deep gratitude for helpful and discerning suggestions, particularly when the book was still in its formative stage. Phyllis Faber, editor of *Fremontia*, the journal of the California Native Plant Society, supported the idea of this book with great enthusiasm from its earliest beginnings and volunteered to go to each and every one of the sites. Phyllis has also been an excellent editor. Harold Mooney, Professor of Biological Sciences, Stanford University, gave me very useful suggestions, especially about the organization of the first several chapters.

Many people have been helpful on my travels in mediterranean climate regions. Adrianna Hoffmann, author of an excellent series of field guides for Chilean flora, was particularly generous with her time and good advice. Professor Gloria Montenegro had valuable suggestions for botanizing in Chile and shared her excellent electron micrographs of sclerophyll leaves. Professor Mary Arroyo was kind to provide information and hospitality. Steven Hopper, Director of the Botanical Garden at Kings Park, Perth, Australia, engaged us in interesting discussions on the origin and diversity of the flora of Western Australia and had many suggestions for our trip through the southwestern part of the state. Bev Overton, author of a well-illustrated book on Kangaroo Island flora, generously guided us in some instructive roadside botanizing. Ian Green, Fiona Dunbar, and Owen Mountford of Greentours, based in Stoke-on-Trent, England, guided us to many of the botanically rich and little visited areas of Central Crete. Many of my photographs from that trip illustrate the chapter on the Mediterranean Basin.

The preparation of the book benefited from the outstanding library resources for botany and plant ecology in the San Francisco Bay Area. Barbara Pitschel, librarian at the Helen Crocker Russell Library of the Strybing Arboretum and Botanical Gardens of San Francisco has been unfailingly helpful in suggesting books in this superb and accessible collection. Annie Malley has provided invaluable assistance at the Biodiversity Resource Center of the California Academy of Sciences in San Francisco. Frank Almeda, Botanist at the California Academy of Sciences in San Francisco warmly introduced me to the Botany Department with its excellent library. Ingrid Radkey, reference librarian of the Bioscience Library at the University of California, Berkeley had the patience and goodwill to get me started on useful data bases for searching the botanical literature.

My family played an essential part in the preparation of this book. My son Tom educated me in the ways of graphic software, provided botanical insights, and gave me many valuable suggestions for the text and graphics. My wife Mary was unfailingly enthusiastic about this project, contributed important editorial suggestions, and provided creative ideas for graphics.

Our weekend place in Sonoma County provided hands-on experience with growing plants from the mediterranean climate regions. I was lucky enough to be a stone's throw from Manning's Heather Farm, a nursery specializing in many mediterranean climate plants. Evie and Gerry Matheson, who ran this nursery for many years, became good neighbors, friends, and teachers. Another outstanding nursery within a few miles is Western Hills, now run by Maggie Wych, who knowledgeably provides for the needs of collectors of mediterranean climate plants and who lent me some of the photographs of Lester Hawkins, a founder of this showplace nursery. Nearby, at Harmony Farms Nursery, David Henry, Pat Parnell, and Robin Rogers have been consistently helpful and informative.

My interest in the mediterranean climate regions was stimulated at Strybing Arboretum and Botanical Gardens in San Francisco where I became a volunteer after I retired from the faculty of the School of Medicine at the University of California in San Francisco. The Arboretum features plant collections from each of the mediterranean climate areas of the world. My familiarity with these was fostered by Bob Hyland, who was the capable and enthusiastic Director of Education when I became a docent-guide. I also learned a great deal from Don Mahoney, Glen Keator, Dick Turner, and my fellow docents who sharpened the group's botanical knowledge and appreciation through weekly "enrichment" tours. Particular thanks also go to my fellow docents Pat Castor and Julie Brook, who went over the manuscript to make helpful corrections and provide suggestions.

Recently, I joined the committee that develops interpretive signs for the Strybing Arboretum. I have learned a great deal from my fellow committee members. Readers of this book who visit the Arboretum will find familiar maps, figures and text on signs in the Chilean Garden and in the soon to be designed Gondwana Circle.

The rich collection of maps, diagrams, and drawings that I was able to select for the book is a tribute to the generosity of many individuals. Excellent drawings and diagrams are included through the courtesy of Stuart Allen of Allen Cartography, Allison Atwill and John Evarts of Cachuma Press, John S. Beard, the Botanic Gardens of Adelaide, Max Debusshe, Fernwood Press, Francesco di Castri (Elsevier, Cambridge University Press), C. E. Conrad, Aljos Farjon, Murray Fagg, Bert Hölldobler and Harvard University Press, Jochen Kummerow, Jack Major, Harold Mooney (Institute of Ecology), Annual Review of Ecology and Systematics, University of California Press, Colin Patterson-Jones, Oliver Rackham, Oxford University Press, Bill Smith and the San Francisco Chronicle, and Springer Verlag. Electronmicrographs are included through the generosity of Gloria Montenegro. Permission to reproduce topographical maps was granted by the publishers Rand McNally and John Wiley and Sons. Most of the photographs are my own, but I was able to fill in important gaps through the kindness of Adrianna Hoffmann, David Cavagnaro, and Maggie Wych.

It has been a special pleasure to work with Beth Hansen-Winter, the designer of this book. Her artistic talents and love of plants have played a major role in giving this book a fresh, lively, and interesting look.

—*Peter R. Dallman*

PREFACE

A few years ago when I was developing a course in biogeography, I planned to focus on comparisons of plant life in the five regions of the world with a mediterranean climate. I had visited all these regions and knew that their floras are very different because of the separate evolutionary origins and that much had been written about the similarities among their vegetation types, a result of convergent evolution. At the same time, nowhere in Western Australia or South Africa had I seen a vegetational landscape that I could mistake for a Californian one. But the vegetation types of central Chile and the western Mediterranean strikingly resemble those of foothills around California's Central Valley. In researching this topic, I found reports that compared aspects of plant life between South Africa and Western Australia, between Chile and California, between the Mediterranean and California, etc., but very few papers compared a single feature of the floras or vegetation of all five mediterranean-climate regions.

In this engaging and beautifully illustrated book, Peter Dallman describes the five regions of the world with a mediterranean climate (their climates are not so similar as I once believed), the diverse adaptations that enable plants to survive the prolonged summer droughts typical of these regions, the plant communities found there, and human influences that have shaped the physical and botanical landscapes. For each region he describes and illustrates significant features of the terrain, environmental influences, and vegetation types. As a traveler, Dallman has first-hand knowledge of these places, has read widely, and has distilled myriad facts into a highly readable and engaging synthesis for those interested in the rich array of plants that grow in these regions. Chapter 10 presents useful suggestions for those planning a trip to any of these regions and recommends books that will enhance their visits.

This book is unique in its coverage and written in a way that makes it easily accessible to the general reader as well as to biologists, geographers, horticulturists, and prospective travelers.

—*Robert Ornduff*

MEDITERRANEAN CLIMATE AREAS

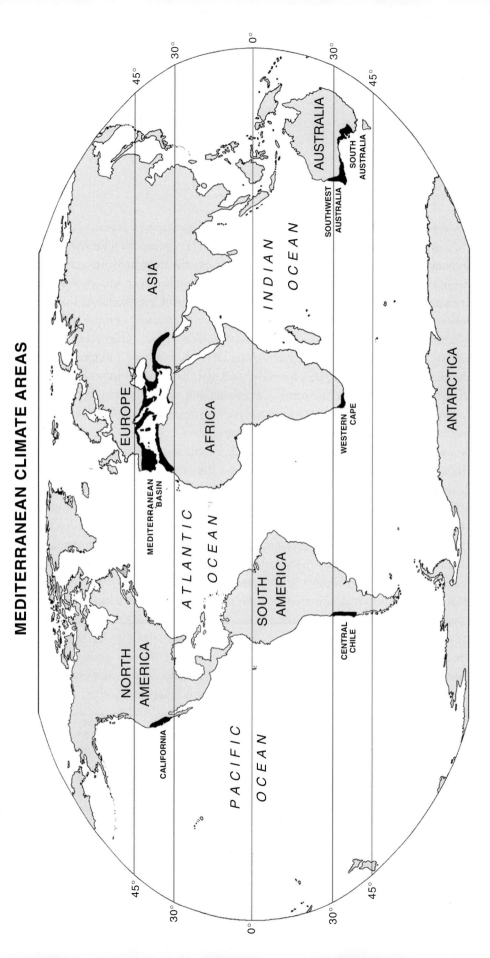

Fig. 0.01. Map of the world with mediterranean climate areas shown in black. All mediterranean climate areas lie between about 30° and 45° of latitude. They are all near the coast on the western edge of continents. (Adapted from DiCastri et al., 1981.)

INTRODUCTION

The region around the Mediterranean Sea has a climate of mild, rainy winters and hot, dry summers. This distinctive climate pattern is also found in four other widely separated parts of the world: California, Chile, South Africa, and Australia. Early European explorers were struck by the similarities in climate and appearance of the vegetation. Subsequently, settlers became successful in growing mediterranean* crops, such as wheat, grapes, olives, citrus fruit, and figs, in each of the distant regions. The shared mediterranean climate pattern of the five areas is now associated with fine wines and excellent fruit. The *native* plant flora of each of these regions, however, is genetically distinctive. Nevertheless, plant species have remarkably many features in common as a result of having survived the challenges of a similar climate.

All five of the world's mediterranean climate regions are concentrated within the latitudes of 30° and 45°, a little less than halfway from the equator to the poles (Fig. 0.01) (DiCastri et al., 1981). The Mediterranean Basin and California are located in the northern hemisphere, while Central Chile, Western Cape Province of South Africa, and the states of Western and South Australia are located in the southern hemisphere. In the southern hemisphere, July is typically the coolest winter month and January the hottest summer month, the reverse of the seasons in the northern hemisphere.

Mediterranean climate areas favor year-round outdoor activity, which fosters an enjoyment of nature and an interest in conservation among natives and visitors alike. This book pulls together information, photographs, illustrations, diagrams, and maps to give an overview of plants, landscape, and climate. Chapters about plant communities of the five mediterranean climate areas are supplemented by a chapter with tips for the traveller. A broader appreciation of the unique natural vegetation of mediterranean climates should lead to stronger support for its conservation.

In ancient times, Aristotle classified climates into three types (James, 1965). He considered Greece and the Mediterranean region as a whole to be temperate and the only zone suitable for civilized life. He regarded the areas to the north as frigid and too stormy and the regions to the south as too torrid for any but barbarians. Only a small fraction of the once abundant vegetation of the Mediterranean Basin remains; however, we are now better informed about the type of plants that thrive in the mediterranean climate. In addition, we

---

* Mediterranean is capitalized when it refers to the Mediterranean Sea and the area around it. The lower case is used when describing the mediterranean climate, mediterranean types of vegetation, and other characteristics of the Mediterranean Basin that also apply elsewhere in the world.

can travel to see distinctive forms of vegetation in the five parts of the world that share this unique climate.

How This Book Is Organized

Chapter 1 describes the mediterranean climate, how it is related to other climates, and the five areas of the world where it is found. The climate has a profound influence on the pattern of plant growth. It is also associated with frequent fires, particularly at the end of the dry season, and this strongly influences the vegetation of mediterranean climate regions.

Chapter 2 examines the ancient origins of mediterranean-type vegetation and the apparent paradox of a mediterranean climate having developed much more recently.

Chapter 3 presents the various ways in which plants of the mediterranean climate are adapted to their environment.

Chapter 4 introduces major plant communities. Each mediterranean climate region has an unusual diversity of plants with a high percentage of rare and endangered species.

Chapters 5 through 9 describe the landscape, climate, and major plant communities of each mediterranean climate region, illustrated with maps, photographs, and drawings of representative species. The evergreen, shrub-dominated chaparral-like forms of vegetation are discussed first. Next, the lower growing and often softer coastal scrub is considered. Woodland and forest are described last. The purpose of these chapters is to illustrate common themes as well as distinctive characteristics of each region. Based on their similarities in vegetation, climate, topography, and soil, California and Chile form a pair and are therefore described in sequence, in Chapters 5 and 6, to facilitate comparisons. Western Cape and Australia are paired for the same reason in Chapters 7 and 8. The Mediterranean Basin is by itself in Chapter 9 because of its complex geography, great size, and plant contributions from three continents, Europe, Africa, and Asia. The Mediterranean Basin has also had the most prolonged, historically documented impact of human settlement.

Chapter 10 has travel suggestions for each of the mediterranean climate areas. Vacations focused on the enjoyment and appreciation of nature are becoming more popular but are largely ignored by most travel books. Wildflowers are at their best in mediterranean climate regions during off-peak travel seasons when planes are less crowded and accommodations are readily available.

In the Bibliography the list of references is generous because the topic has, to my knowledge, not previously been covered for the general reader. This book benefits from a growing interest in the mediterranean climate areas among ecologists and botanists. Symposium volumes dealing with subjects such as plant adaptations, soil nutrients, the role of fire, and species diversity in the mediterranean climate regions are appearing with increasing frequency.

FIG. 1.01. *Enjoying balmy, short-sleeve weather in early spring; coastal scenery in southwestern Australia.*

1.
THE MEDITERRANEAN CLIMATE

THE WEATHER PATTERN AND WHERE TO FIND IT

T*he mediterranean climate regions found in California, Chile, South Africa, Australia,* and the Mediterranean Basin share a climate of dry summers with brilliant sunshine and clear blue skies. Rainfall is concentrated during the mild, winter half of the year and averages roughly10-40 inches (25-100 cm) annually. Snow is rare except at high elevations. The total amount of winter rain is highly variable from year to year, coming in storms that may last a few days. Between storms, the weather is often crisp and clear. Summer rains are rare and scant.

All five mediterranean climate regions are located on the western or southwestern coasts of continents (Fig. 0.01, page x), where there are typically cold, offshore ocean currents. The influence of large oceans with an upwelling of cold water off the coast has a moderating effect on summer temperatures. Cold offshore currents include the California current off the west coast of the United States, the Humboldt current to the west of Chile, the Benguela current near the west coast of South Africa, and the Canaries current off Morocco and Portugal. Southwestern Australia seems to be an exception because the southward-flowing offshore Leeuwin current is warm (Hopper et al., 1996). The Mediterranean Sea, because it is

enclosed by land, warms up considerably more during the summer than does the Atlantic Ocean to the west. Consequently, lands that border the Mediterranean Sea have hotter summers than mediterranean climate regions that face an ocean with cold offshore currents.

A mediterranean climate occurs on about two percent of the world's total land area (Thrower and Bradbury, 1973; World Conservation Monitoring Centre, 1992). The largest region of land with this climate is the Mediterranean Basin with about 60 percent of the world's total mediterranean climate area (Fig. 1.02). The Mediterranean Sea splits the basin up into several geographically distinct regions (Fig. 1.03) and over 15 countries. The Mediter-

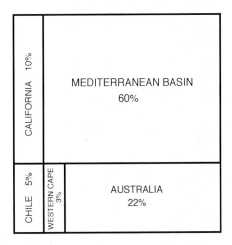

FIG. *1.02. Relative areas of those parts of the world that have a mediterranean climate. (From DiCastri, 1881.)*

1

ranean Basin is the only mediterranean climate region that includes parts of three continents, giving it a very rich flora particularly where continents meet.

The next largest area of mediterranean climate, with about 22 percent of the total, is found in Australia. It is split into two parts that are separated by about 500 miles (800 km) of intervening desert and semi-arid terrain: the southwestern part of the state of Western Australia lies to the west and the southern part of the state of South Australia to the east. The mediterranean climate area of South Australia extends eastward for a short distance into the states of Victoria and New South Wales.

Among the remaining three regions, California occupies 10 percent of the world's total mediterranean climate area, comprising most of the state west of the Sierra Nevada together with a small adjacent portion of the Mexican state of Baja California Norte. Central Chile accounts for about five percent of the world's total. Lastly, the mediterranean climate area of Western Cape is by far the smallest, occupying only three percent, but it has the highest concentration of plant species.

FIG. 1.03. *Mediterranean climate areas shown at the same scale. California and the Mediterranean Basin, the two northern hemisphere areas, are shown above. The three southern hemisphere areas, Central Chile, the southwestern portion of Western Cape (South Africa), and the mediterranean climate areas of Western and South Australia, are below. High mountains are present only in California, Chile, and the Mediterranean Basin. California and Chile both have long central valleys with lower mountains and hills along the ocean to the west. The mediterranean climate areas in Western Cape Province and Australia consist only of tablelands, plains, and lower mountains. (Thrower and Bradbury, 1977.)*

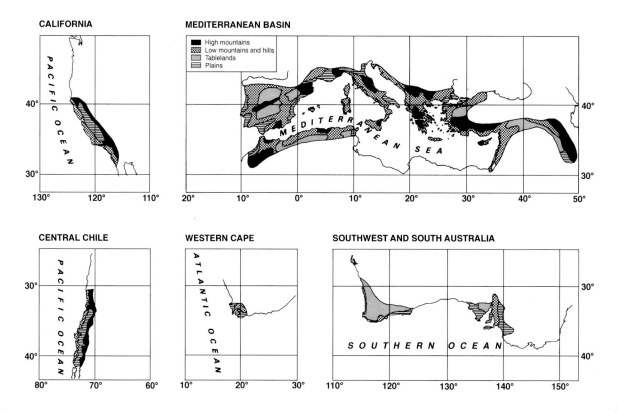

FIG. 1.04. *Summer coastal fog at Año Nuevo State Park, south of San Francisco. Monterey pines* (Pinus radiata) *grow on the top of the cliffs.*

Mediterranean-type Vegetation Beyond Mediterranean Climate Boundaries. Vegetation typical of the mediterranean climate may extend beyond that climate into both drier and moister areas. In some dry coastal areas, for example, summer temperatures are moderated by fog (Fig. 1.04), which provides additional precipitation from fog drip. Fog is common along the coasts with cold offshore currents, especially in California and Chile and on the western coast of the Western Cape Province. Moist fog in the form of fog drip can add as much as 10 inches (25 cm) of precipitation per year, beyond what is recorded in weather records, supporting plants that require extra moisture.

Other plant species of mediterranean climate regions extend into moister adjacent areas, where summers are still the driest season but where drought is generally brief and mild. Such areas include coastal northern California and the western parts of Oregon and Washington, the lake district of Chile, and northern Spain, Italy, and Greece.

Population Pressures. A rough approximation of area, population, and population density (people per square mile) for each of the mediterranean climate areas is shown on Table 1.01. Population densities are greatest for the Mediterranean Basin; intermediate for California, Chile, and the Western Cape; and lowest for Australia. A low population density improves the prospects for maintaining large areas of native vegetation in a relatively unspoiled state. The differences in population density help to explain why conservation efforts have been far more successful in Australia than in the Mediterranean Basin.

TABLE 1.01. ESTIMATED AREA AND POPULATION OF THE
MEDITERRANEAN CLIMATE REGIONS

Region	Area, miles^{2*}	Population*	Population Density/mile^2
California	120,000	27,000,000	225
Chile	58,000	11,000,000	190
South Africa	27,000	3,400,000	125
Australia	200,000	3,000,000	15
Mediterranean Basin	500,000	150,000,000	300

*Mediterranean climate areas only, not the entire region.

HOW LOCATION AND TERRAIN INFLUENCE THE CLIMATE

Rainfall and temperature patterns differ considerably within each of the mediterranean climate areas, owing to factors such as latitude, exposure to the ocean, elevation, and orientation to the sun.

Latitude. Mediterranean climate areas experience decreasing rainfall as the latitude approaches 30°, toward the equator. Correspondingly, there is a gradual change in vegetation toward plants that are adapted to semi-arid and desert conditions. In contrast, as the latitude approaches 40° to 45°, toward the poles, rainfall steadily increases and becomes distributed more evenly throughout the year. Beyond 40° to 45°, high levels of rainfall support temperate rainforests in the Pacific northwest of the United States and to the north of the Mediterranean Basin in Europe. Similarly, there are rainforests to the south of the mediterranean climate area of Central Chile. Western Cape and Australia have only ocean in the polar direction and lack an analogous, adjoining area of rainforest.

Exposure to the Ocean. Land temperatures change more rapidly than ocean temperatures during the daily cycles of light and dark and the annual cycles of summer and winter. Temperatures of coastal land areas are moderated by oceans, decreasing the temperature differential between day and night and between winter and summer relative to adjacent inland regions. Mediterranean climate regions are primarily coastal and are described as having a maritime climate. In contrast, continental climates with extremes of temperatures are found far from the influence of the ocean. In an intermediate position are the more inland mediterranean climate areas that combine some degree of coastal and continental influences. The differences between maritime and continental climates are summarized as follows (Ornduff, 1974):

MARITIME CLIMATE	CONTINENTAL CLIMATE
Coastal location	Inland location
Winters cool	Winters cold
Summers warm	Summers hot
Small daily temp. range	Large daily temp. range
Small seasonal temp. range	Large seasonal temp. range
High relative humidities	Low relative humidities

In the mediterranean climate area of California, the maritime influence is great not only near the coast, but beyond gaps in the coastal ranges, such as in the wine-growing Napa and Sonoma valleys that open onto San Francisco Bay and the Golden Gate. Continental influences increase further inland in the Central Valley and the foothills of the Sierra Nevada.

Elevation and Terrain. Elevation is important in relation to temperature and rainfall. Temperature in the mountains decreases by roughly 2.8°F per 1000 feet (5°C/1000 m). Rainfall is typically greater with increased elevation, particularly on western slopes that face the ocean and prevailing winds. Even hills of moderate size can be much rainier than nearby, lower-lying areas. For example in Adelaide, the capital of South Australia, the center of the city, which is near sea level, has only 21 inches (53 cm) of rain per year, while the nearby suburb of Stirling has more than twice as much by virtue of its location near the peak of the 2400 foot (727 m) Mt Lofty. Vegetation correspondingly changes with elevation. The effects of elevation and terrain on vegetation are graphically illustrated on schematic transects through California, shown in Fig. 5.05 on page 64-65.

The configuration of coastal mountains can create contrasting microclimates within a small area. This is particularly evident in the San Francisco Bay Area, where breaks in the coastal range allow cool coastal fog to penetrate inland. Dense, moist coastal redwood forests grow in close proximity to much drier chaparral and oak woodlands. Mountain chains that parallel the coast create rain shadows, giving interior valleys a drier climate. Berkeley, on San Francisco Bay, is much rainier than the suburb of Walnut Creek, which lies just beyond a range of hills to the east. Similarly marked variations in microclimate are found around Cape Town in South Africa, where opposing climatic influences from the Atlantic and Indian Oceans meet around Table Mountain.

All of the regions with mediterranean climates are shown together at the same scale in Fig. 1.03. California and the Mediterranean Basin, which are in the northern hemisphere, are above, and the three southern hemisphere zones, Central Chile, Western Cape Province, South Africa, and Southwestern and South Australia, are below. High mountains, shown in black, are present only in California, Chile, and the Mediterranean Basin. These three areas have high, sharp, faulted mountains that are geologically young and that were partly glaciated during recurrent ice ages of the last three million years. In these three areas, soils tend to be richer in nutrients than in mediterranean climate areas of South Africa and Australia (Chapter 3, page 39).

The Mediterranean Basin, California, and Chile all have winter landscapes characterized by green valleys with snow-covered mountains in the distance (Fig. 1.06). California and Chile resemble one another most closely in topography. Both have long central valleys running north to south that are flanked by high snow-covered mountains to the east and lower coastal mountains and the Pacific Ocean to the west.

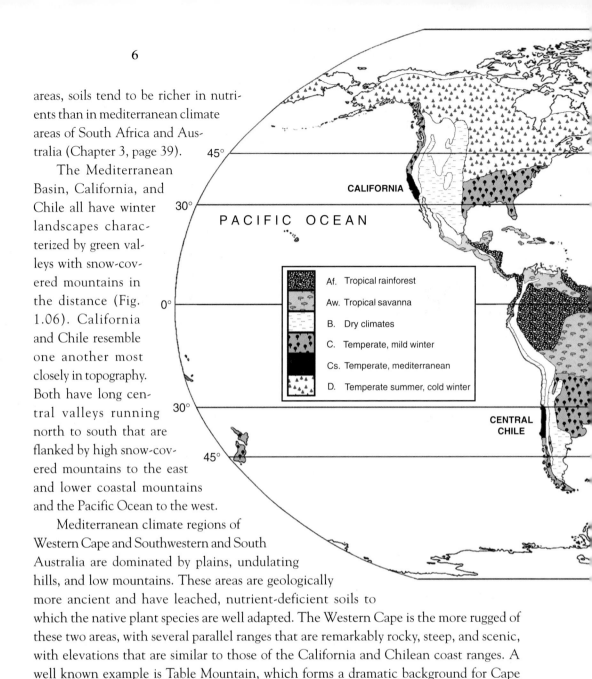

Mediterranean climate regions of Western Cape and Southwestern and South Australia are dominated by plains, undulating hills, and low mountains. These areas are geologically more ancient and have leached, nutrient-deficient soils to which the native plant species are well adapted. The Western Cape is the more rugged of these two areas, with several parallel ranges that are remarkably rocky, steep, and scenic, with elevations that are similar to those of the California and Chilean coast ranges. A well known example is Table Mountain, which forms a dramatic background for Cape Town.

Orientation of Mountain Slopes. In mountainous areas, the orientation of slopes has a profound influence on vegetation. South-facing slopes are sunny and dry in California and the Mediterranean Basin but are shady and moist south of the equator in Chile, Western Cape, and Australia.

RELATIONSHIP TO OTHER CLIMATES

The simplest and most widely used system for classifying climates was devised by Wladimir Köppen, a Russian-born climatologist and botanist. Working in Germany during

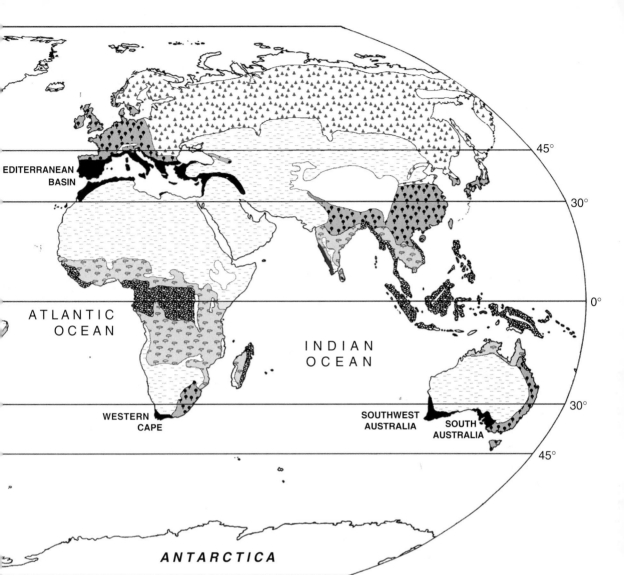

FIG. 1.05. *Major climate zones of the world.*

the 1920s and 30s, he established quantitative temperature and precipitation criteria for various climates that roughly corresponded to their vegetation.

Köppen's climate zones are named using capital and small letters. The six major climate zones are A, B, C, D, E, and H. The first four of these are presented in Fig. 1.05. The mediterranean climate zones are shown in black and fall within category C, the temperate climate with mild winters.

The six major climate zones listed in alphabetical sequence are as follows:

A. Tropical climates are rainy for a substantial part of the year. Examples are tropical rain forests of the Amazon and Congo river basins (Af.) and drier tropical savannas of sub-Sahara and East Africa (Aw.).

B. Dry climates include deserts and semi-arid steppes of the southwestern United States, northern Mexico, the Sahara, and most of the Middle East, Central Asia, and the interior of Australia.

C. Temperate, mild winter climates include the mediterranean climate regions that are examined in this book. Also in category C are areas of year-round precipitation, such as the Pacific Northwest, southern Brazil and northeast Argentina, Great Britain, Germany, most of France and the surrounding parts of northern Europe, northern India, southern China and Japan, and New Zealand and eastern, coastal Australia. In all of these areas, the coldest winter month is mild with an average temperature above 32°F (zero°C) but below 64°F (18°C). The temperature of the warmest summer month averages more than 50°F (10°C).

D. Temperate summers but cold winter climates include the great forests of Canada, Alaska, the northeastern U.S., Scandinavia, and Russia, all of which are covered with snow in the winter.

E. Polar climates have no month with an average temperature above 50°F (10°C), and there is little or no vegetation. They are found on the polar ice caps and the tundra of northernmost Alaska, Canada, Greenland, and Russia.

H. Highland climates are represented near and above the tree line in the Sierra, Rockies, Andes, Alps, and Himalayas. They are areas with scant vegetation, where the climate changes substantially with elevation. The climate zones E and H are shown in white to convey their paucity of vegetation.

TYPES OF MEDITERRANEAN CLIMATES

Mediterranean climates that have wet winters and dry summers are a subcategory of **C. Mild winter climates** and are designated as Cs, the s signifying that summer is the dry season. Cs climates are further divided according to differences in summer weather (James, 1965). In California, for example, three subdivisions are:

Csa. Climate with hot summers, the highest monthly temperature averaging over 72°F (22°C), includes Sacramento, Stockton, and Los Angeles.

Csb. Climate with moderate to warm summers, the highest monthly temperature averaging below 72°F (22°C), includes Santa Rosa, Santa Barbara, and Long Beach.

Csbn. Coastal areas that have cool summers with frequent fogs, are exemplified by San Francisco, Santa Cruz, and Monterey (n stands for Nebel, the German word for fog).

Duration of Summer Drought. The mediterranean climate and adjacent areas can be described on the basis of the average duration of summer drought (Emberger, 1955, Bagnouls and Gaussen, 1957; DiCastri, 1981; Nahal, 1981).

11 to 12 dry months: very arid	5 to 6 dry months: subhumid
9 to 10 dry months: arid	3 to 4 dry months: humid
7 to 8 dry months: semi arid	1 to 2 dry months: very humid

These categories correspond reasonably well with differences in vegetation. They are

TABLE 1.02. RAINFALL IN THE WETTEST AND DRIEST MONTHS IN CITIES WITH MEDITERRANEAN CLIMATES

Region and City	Rainfall, cm.		
	Annual	Wettest mo.	Driest mo.
California			
San Francisco	53	11.6	0.0
Los Angeles	37	7.8	0.0
San Diego	26	5.5	0.0
Sacramento	53	5.3	0.0
Chile			
Santiago	36	8.5	0.2
Valparaiso	46	12.8	0.2
Concepción	129	23.8	1.7
La Serena	13	4.4	0.0
Western Cape			
Cape Town	51	8.5	0.8
Australia			
Adelaide	52	6.6	2.3
Perth	89	19.2	0.7
Albany	101	15.2	3.1
Geraldton	46	11.3	0.5
Mediterranean			
Casablanca	51	12.5	0.0
Lisbon	71	11.1	0.3
Barcelona	59	8.0*	2.9
Marseille	55	7.6*	1.1
Genoa	130	18.1*	4.9
Tunis	46	7.0	0.1
Rome	87	12.8*	1.4
Split	82	13.0	4.0
Athens	40	7.1	0.6
Izmir	65	12.2	0.5
Haifa	67	18.5	0.2

*October is the wettest month; December, January or February is in all other northern hemisphere cities and May, June or July in all southern hemisphere cities.

(Note: all of these cities are shown on the maps in Chapters 5 through 9)

(Adapted from Rother, 1984.)

particularly suitable for the diverse Mediterranean Basin itself, because, unlike the other areas, its season of peak rainfall varies substantially from one region to another. These categories also illustrate the climate of Chile in Fig. 6.04 (page 94).

FREQUENT WILDFIRES

The occurrence of drought during the hottest part of the year increases the flammability of dead wood. This is why regions with a mediterranean climate are more susceptible to fire than other parts of the world. Autumn, winter, and spring rains support the growth of dense shrubs and trees, which produce an increasing amount of dry fuel over the years. Most fires occur in the summer and early autumn. They are particularly destructive where there has been a prolonged buildup of fuel. Ironically, this is often where fire prevention has been practiced most successfully and for the longest period (Keeley and Keeley, 1986).

All mediterranean climate regions are relatively dry in the summer, but some areas are drier than others. Table 1.02 shows the rainfall in selected cities, all of which are shown on the maps that follow. The cities listed for California and Chile (other than Concepción) experience almost no rainfall during the driest summer month. In contrast, Cape Town, South Africa, the Australian cities, and most Mediterranean Basin cities have occasional summer rains. These light and infrequent rain storms have a moderating effect on the

Fig. 1.06. *Snow-capped Mt. Psiloritis (Mt. Ida) dominates this springtime view of the Amari Valley in Crete. Sclerophyll trees and shrubs in the foothills behind the vineyard remain green during the winter.*

TABLE 1.03. AVERAGE TEMPERATURE OF THE HOTTEST AND COOLEST MONTHS AND LATITUDES OF CITIES WITH MEDITERRANEAN CLIMATES

Region and City	Temperature, °C			Latitude*
	Annual	Coolest mo.	Hottest mo.	degrees
California				
San Francisco	13.8	10.4	16.7[+]	38.0
Los Angeles	18.0	13.2	22.8	34.0
San Diego	17.2	13.1	21.5	33.5
Sacramento	15.3	7.6	22.4	38.5
Chile				
Santiago	14.0	8.1	20.0	33.5
Valparaiso	14.7	11.8	18.0	33.0
Concepción	13.0	9.1	18.0	37.0
La Serena	14.7	11.7	18.4	30.0
Western Cape				
Cape Town	17.0	12.6	21.5	34.0
Australia				
Adelaide	16.7	11.2	22.6	35.0
Perth	18.1	13.1	23.9	32.0
Albany	15.7	12.1	19.4	35.0
Geralton	19.8	15.4	24.1	29.0
Mediterranean				
Casablanca	17.5	12.4	23.0	33.5
Lisbon	16.6	10.8	22.2	38.5
Barcelona	16.4	9.4	24.4	41.5
Marseilles	14.4	5.5	23.3	43.5
Genoa	15.6	7.9	24.1	44.5
Tunis	17.7	10.2	26.2	37.0
Rome	15.5	6.9	24.7	42.0
Split	16.1	7.8	25.6	43.5
Athens	17.8	9.3	27.6	38.0
Izmir	17.4	8.3	27.0	38.5
Haifa	21.4	13.9	28.3	33.0

*To the nearest 0.5°

[+]September is the hottest month; July or August is in all other northern hemisphere cities and January or February is in all southern hemisphere cities.

(Adapted from Rother, 1984.)

intensity of summer drought. However, they may also precipitate wildfires when they are accompanied by lightning.

The moderating maritime influence of most mediterranean climate areas gives them mild winters with average temperatures well above the freezing point even during the coldest month. These conditions support growth of broadleaf evergreen shrubs and trees, giving a lush appearance to the winter landscape. The mildest temperatures are close to the coast. Further inland and at higher elevations, climate becomes more continental with greater daily and seasonal temperature variations. Average temperature during the coolest

FIG. 1.07. *Climate charts for cities in each of the mediterranean climate areas: San Francisco, California; Athens, Greece; Valparaiso, Chile; Cape Town, South Africa; and Perth, Australia. The horizontal axis for the southern hemisphere cities shows the months starting with July rather than January, giving a similar pattern to all of the charts, with summer in the middle, irrespective of the reversal of the seasons.*

The thick part of the horizontal axis shows the months of water deficit (dry months), when the precipitation curve falls below the temperature curve. In San Francisco, for example, the precipitation curve first falls below the temperature curve during April and does not rise above it again until mid October, a duration of 6.5 months. Among the other cities, Valparaiso has the most prolonged water deficit with a duration of 7.5 months; Cape Town and Perth have the shortest periods of just over 5 months.

The magnitude of water deficit is greatest in Athens, owing to the combined effects of low winter rains and high summer temperatures. Plants native to the Athens area must therefore be particularly well adapted to drought. (From Major, 1988.)

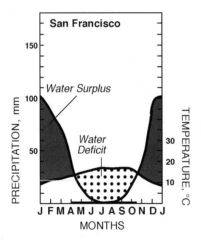

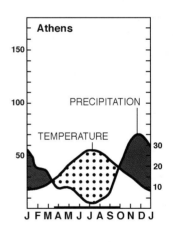

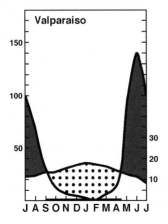

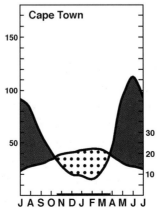

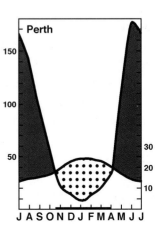

and hottest months (data from Rother, 1984) and latitudes of selected cities with mediterranean climates are shown in Table 1.03.

WEATHER CONDITIONS IN RELATION TO PLANT GROWTH

Summer drought is the most severe stress for plants in mediterranean climate areas. Both the duration of dry weather and the intensity of summer heat affect plant survival. The more intense and prolonged the heat, the more severe the drought stress. A good way to appreciate the relationship between rainfall and temperature is on a graph of monthly averages for precipitation and temperature. The scales that are commonly used for this purpose show precipitation in millimeters at twice the numerical scale for temperature in °C. There is generally adequate moisture for plant growth when the curve for precipitation is above the temperature curve and a water deficit or dry conditions exist for plants when it falls below the temperature curve (Bagnouls and Gaussen, 1959). Fig. 1.07 shows such plots for a city in each of the mediterranean climate areas: San Francisco, California; Athens, Greece; Valparaiso, Chile; Cape Town, South Africa; and Perth, Australia (Major, 1988).

In addition to showing duration of dry periods, different patterns of rainfall and temperature in each of the cities are apparent in Fig. 1.07. Perth has the heaviest winter rains. Consequently, there can be ample soil water during the early part of the dry season to support plant growth. Furthermore, Perth, like Cape Town, also has some summer rain. Athens has the lightest winter rains combined with the highest summer temperatures, conditions that aggravate summer drought. San Francisco and Valparaiso are more completely dry during the summer than the other cities. However, this severe dryness is counterbalanced by more moderate summer temperatures and by fog.

UNUSUAL WEATHER

Information presented so far has been based on average weather data. Such figures are useful in comparing one region to another, but they misleadingly ignore deviations from the average that occur with drought, heat waves, and severe frost. Rainfall is notoriously variable from year to year, and this variability is greatest where average annual rainfall is lowest. Plant species in mediterranean climate areas are adapted to survive these conditions. In fact, atypical conditions that threaten plant survival are among the forces that drive evolution. Species that succumb to extreme conditions make room for the diversification of species whose characteristics favor their survival.

It is revealing to examine day-by-day weather records for any given year. Fig. 1.08, for example, is a schematic representation of 1993 rainfall and temperatures in San Francisco, a city notable for its unusually moderate climate. Even though annual averages for 1993 were close to those for previous years, most individual months were extreme in some respect that affected plant growth. January, February, and April were unusually wet. Record-breaking heat was recorded in June. The entire months of June, July, August, and October were unusually hot, setting up the conditions for devastating fire storms. December was

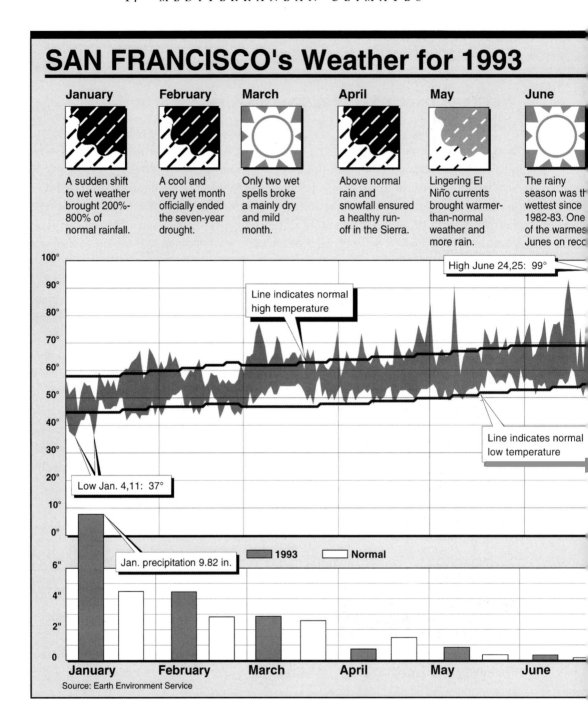

SAN FRANCISCO's Weather for 1993

January
A sudden shift to wet weather brought 200%-800% of normal rainfall.

February
A cool and very wet month officially ended the seven-year drought.

March
Only two wet spells broke a mainly dry and mild month.

April
Above normal rain and snowfall ensured a healthy run-off in the Sierra.

May
Lingering El Niño currents brought warmer-than-normal weather and more rain.

June
The rainy season was th wettest since 1982-83. One of the warmes Junes on reco

High June 24,25: 99°

Line indicates normal high temperature

Line indicates normal low temperature

Low Jan. 4,11: 37°

Jan. precipitation 9.82 in.

■ 1993   ☐ Normal

January   February   March   April   May   June

Source: Earth Environment Service

FIG. 1.08. *Expect atypical weather. San Francisco's weather for 1993 was average in annual precipitation (in inches) and temperature (°F). Nevertheless, almost every month had some very atypical characteristics that affected plant growth. Although plant growth was stimulated by unusually heavy January and*

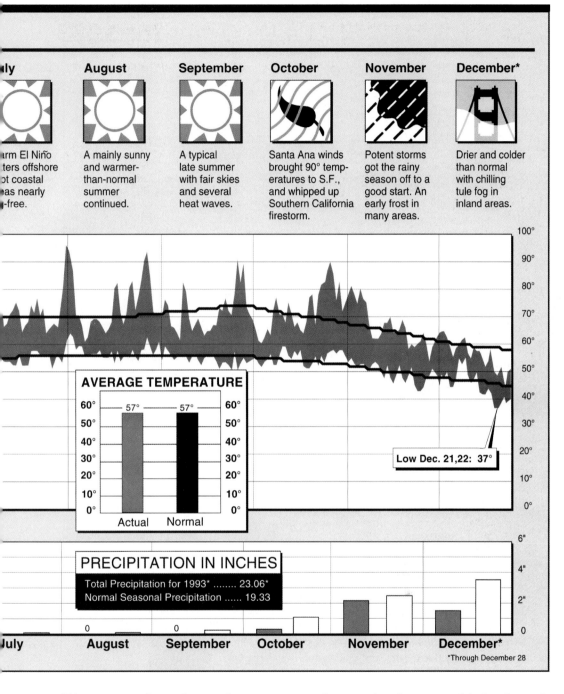

July	**August**	**September**	**October**	**November**	**December***
...rm El Niño ...ters offshore ...t coastal ...as nearly ...-free.	A mainly sunny and warmer-than-normal summer continued.	A typical late summer with fair skies and several heat waves.	Santa Ana winds brought 90° temperatures to S.F., and whipped up Southern California firestorm.	Potent storms got the rainy season off to a good start. An early frost in many areas.	Drier and colder than normal with chilling tule fog in inland areas.

AVERAGE TEMPERATURE

Actual 57° — Normal 57°

Low Dec. 21,22: 37°

PRECIPITATION IN INCHES

Total Precipitation for 1993* 23.06*
Normal Seasonal Precipitation 19.33

*Through December 28

February rains, a hotter than usual summer constituted a severe drought stress, and the hot, dry winds of October set the scene for severe chaparral fires in the Los Angeles area. (Reproduced with permission of the artist, Bill Smith and the San Francisco Chronicle, 1/2/94.)

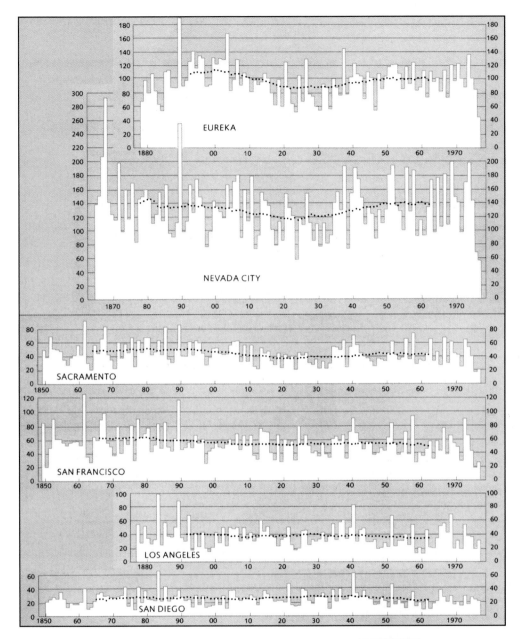

F_IG. 1.09. *Variability in the rainfall of Eureka, Nevada City, Sacramento, San Francisco, Los Angeles, and San Diego during the period of recent history for which weather records are available. The vertical scales are in centimeters.*

The dotted lines represent the 30-year running mean, with each dot representing the average of the annual precipitation from 15 years before to 15 years after. The running average has scarcely changed. However, the year-to-year differences are substantial. Often, there are two to three dry or wet years in a row when rainfall deviates from the running mean by more than 25 percent. (Reproduced with permission from Donley et al., 1979.)

unusually dry. Such variations help to explain why certain types of plants may thrive one year and barely survive or succumb in another.

Plant survival in a mediterranean climate is threatened especially by periods of drought lasting several years. Drought not only impairs plant development and viability, it also increases the likelihood of wildfires and their potential for eliminating some species while stimulating the spread of others. Variability in annual rainfall and the occurrence of drought are shown in weather records from six California cities (Fig. 1.09) (Donley et al., 1979).

Using the Climate Maps to Compare Mediterranean Regions

The climate maps that introduce Chapters 5 through 9 show average annual rainfall for each region, using a similar color scale throughout, to facilitate comparisons. Yellow and light green together can be considered the most typical of the mediterranean climate. Yellow represents the drier end of the mediterranean climate region, where chaparral and other scrub vegetation are primarily found. Areas shown in light green are moister and support evergreen woodland and forest. Outside of the mediterranean area proper, more arid regions with less than about 10 inches (25 cm) of precipitation per year correspond to desert and semi-desert and are shown in orange. Wetter areas receiving more than 40 inches (100 cm) are blue-green and are generally forested.

The temperature maps also use a similar color scale throughout, going from deep green for the coolest temperature, through pale green, deep yellow, orange, and red with increasing heat. Slight differences in temperature ranges are due to the necessary rounding out to match metric and non-metric data. Maps for the average temperature during the coolest month and the hottest month are shown together to allow comparison (in most cases, the maps are for January and July, with January being the hottest month in the southern hemisphere). In both the northern and southern hemispheres, it is customary to refer to the hot season as summer and the cool season as winter. Because of the scale of the maps, the importance of microclimates cannot be shown.

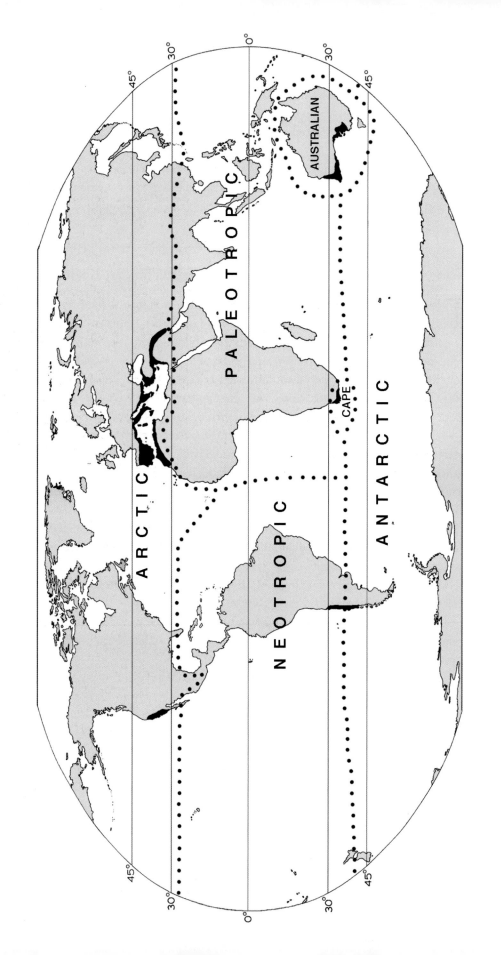

2.
PLANT AND CLIMATE ORIGINS

FLORISTIC KINGDOMS

The plant communities of the world have geographic boundaries that are governed by climate, soil type, and separation of regions by ocean and desert barriers. In addition, the movement of continents by plate tectonics and the changes in climate over millions of years provide explanations for how plants are geographically grouped today.

Plant geographers now divide the world into a hierarchy of kingdoms, regions, and provinces based on the association of related plants (Takhtajan, 1986). This concept was first proposed in a book on plant geography published by the Danish botanist J. F. Schouw in 1822, during the golden age of plant exploration. European botanists who were then collecting new plant specimens in South Africa, Australia, and South America, all in the southern hemisphere, found the flora of these regions to be highly novel and distinctive (Good, 1964). For the most part, plants were too unfamiliar to identify and required both naming and classification. The same applied to the tropical flora of the New and the Old World, which differed not only from each other but also from the vegetation of the temperate climates. The most familiar trees and shrubs for plant explorers from Europe were in North America and East Asia, where many genera and families were the same, although species were different. Such plant associations are the basis for the current division of the world into six floristic kingdoms, of which four include areas of mediterranean climate vegetation (Fig. 2.01). Although there is much overlap from one kingdom to another, each supports a characteristic flora with many endemic species.

The **Arctic Kingdom** is the largest of the six and includes the entire temperate regions of North America and Eurasia. Both California and the Mediterranean Basin are within this kingdom, and they share many genera, such as oak (*Quercus*) and pine (*Pinus*). The madrone (*Arbutus menziesii*) of California and the strawberry tree (*A. unedo*) of the Mediterranean Basin are other prominent examples of a shared genus, as are the Monterey cypress (*Cupressus macrocarpa*) and the Italian cypress (*C. sempervirens*). California and the Mediterranean Basin have the most closely related floras among the five mediterranean climate regions (Raven, 1973).

FIG. *2.01. Floristic kingdoms, representing regions with distinctive floras. (Map based on Takhtajan, 1986.)*

The **Antarctic Kingdom** comprises the southern tip of South America (including the southern part of Central Chile) and far off New Zealand. Forests of southern beech (*Nothofagus*) are found in these distant lands, which are now on opposites sides of the Pacific Ocean but were once connected by a temperate Antarctic continent.

The **Australian Kingdom** is more compact and is restricted to that island continent, with genera such as *Eucalyptus, Acacia, Banksia,* and *Grevillea* found throughout much of the country, despite the difference in pattern of rainfall from one region to another.

The **Cape Floral Kingdom** is by far the smallest of the six kingdoms and includes little more than the minute mediterranean vegetation area of the Western Cape Province. Its flora is distinctive and different from the remainder of the continent, with an enormous diversity in the genera *Protea* and *Erica* and the reed or rush-like family of Restionaceae.

The two remaining kingdoms do not include mediterranean climate areas. They are both tropical. The Paleotropic Kingdom makes up most of the African and Asian tropics and the Neotropic Kingdom comprises the western hemisphere tropics.

These groupings of related plants are clearly not absolute and there are no sharp boundaries between kingdoms. Nevertheless, the world distribution of families and genera of native plants now makes more sense than it once did, in the light of newer information about plate tectonics and continental movement and the geologic and fossil evidence for plant origins.

EARLY ORIGINS OF DROUGHT-RESISTANT, SCLEROPHYLL PLANTS

The typical mediterranean climate pattern of mild, rainy winters and dry summers has developed only within the past three million years (Fig. 2.02). Paradoxically, many of the thick and leathery-leaved, drought-resistant, sclerophyll plants that now grow in the mediterranean climate regions are recognizable in fossil remains from over 40 million years ago, when the climate was hotter and rainfall had a summer maximum (Raven, 1973; Axelrod, 1973 and 1989; Barlow, 1994; Dodson, 1994) (Fig. 2.03). Rainforests and swamps were widespread in and near areas that now have a mediterranean climate. It was temperate and humid even near the poles, which were ice-free (Deacon, 1983; van Andel, 1994).

Evidence from deep sea drilling and from plant fossils indicates that temperatures dropped over the last 40 million years (Axelrod, 1973 and 1989). During this period, substantial year-round rainfall persisted, but its summer peak gradually shifted and eventually

FIG. 2.02. *From early flowering plants to an early botanist: a time line for the history of flowering plants and the mediterranean climate. Of the geological ages, the Cretaceous (144 to 66 million years ago) is the age of the dinosaurs and the emergence of flowering plants. The Tertiary (66 to 1.6 million years ago) is the period during which modern plants and animals diversified. The Quaternary is the age of man. It includes the more recent ice ages, and takes us from 1.6 million years ago to the present.*

Time Line

CRETACEOUS

YEARS AGO
(log scale)

100,000,000 —
- Early flowering plants
- Separation of North America and Eurasia
- Flowering plants resembling modern species
- Dinosaur extinction ends <u>Cretaceous</u> Period

50,000,000 —
- Separation of South America, Antarctica, and Australia
- **Sclerophyll plants** resembling modern species
- Warm, moist climates; no polar ice

- Climate gradually cooling near the poles

10,000,000 —
- Development of the Antarctic ice sheet

5,000,000 —
- Mediterranean salinity crisis

- **Start of the mediterranean climate** at end of the <u>Tertiary</u> and the beginning of the <u>Quaternary</u> Period: the age of man

- Glacial periods of about 100,000 years, separated by warmer, interglacial intervals of about 10,000 years

1,000,000 —

500,000 —
- Early man: Homo erectus

- Alternating periods of cooling and warming result in loss, diversification, and migration of plant species

100,000 —
- Plant species with effective mechanisms for seed dispersal are at an advantage

50,000 —

- Cro-Magnon man

- Cave paintings in Spain and France

- Last glacial maximum (18,000 years ago)

10,000 —
- Domestication of animals

- Start of agriculture and weed dispersal

5,000 —
- Early written records

- Theophrastus, Greek botanist (372-287 BC) Author of Enquiry into Plants (Peri Phyton Historias)

1,000 —

T
E
R
T
I
A
R
Y

Q
U
A
T
E
R
N
A
R
Y

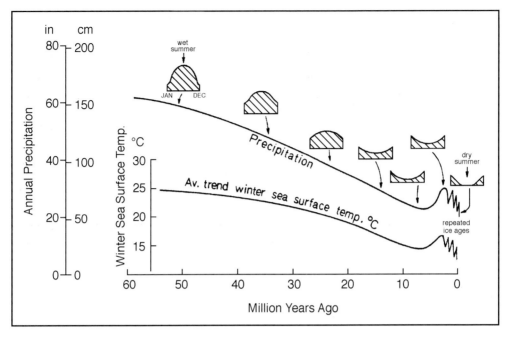

Fig. 2.03. *The mediterranean climate in California originated recently. Fifty million years ago, winter temperatures at the ocean surface were much warmer than now. Annual precipitation was abundant, with a peak in the summer, as shown in the small diagrams above the precipitation curve. The striped areas in these diagrams show the pattern of precipitation from January at the left to December on the right. Wet summers lasted until about 25 million years ago. More recently, as the climate cooled, the summers became the drier parts of the year.*

During the last three million years, there were about 30 ice ages, represented by the zigzag fluctuations in temperature and precipitation. The ice ages occurred with a periodicity of roughly 100,000 years with intervening warmer periods of 10,000 years. It is during these intervening warm periods, one of which we are experiencing now, that the mediterranean climate evolved, becoming progressively drier in the summer. (Adapted from Axelrod, 1973, with permission of the author and publisher.)

reversed to the pattern of winter rain and summer drought that is now found in mediterranean climate regions.

If sclerophyll plants did not initially evolve in a mediterranean climate, how can we explain their earlier origins? The widely accepted conclusion is that ancestors of sclerophyll plants now living in mediterranean climates were preadapted. That is, they originally had characteristics that allowed them to survive and diversify when the climate with dry summers and wet winters eventually developed.

Northern Hemisphere. In the western part of the United States and in Mexico, many of the thick-leaved, evergreen ancestors of present-day chaparral shrubs were found as long as 46 million years ago in the understory of moist broadleaf evergreen forests and woodlands (Raven, 1973; Axelrod, 1989). Subsequently, with an increasingly drier climate, more frequent wildfires, and the growth of high western mountain chains, chaparral plant species were able to spread and diversify under conditions that were too hostile for many tree species under which shrubs had originated. As the climate developed into mediterranean patterns, these preadapted species of shrubs evolved into large areas of chaparral vegetation. In addition, many drought-tolerant genera of shrubs and trees of mediterranean California continue to thrive in the semi-arid highland areas of Arizona and northern Mexico where there is summer rainfall. Here are found manzanita (*Arctostaphylos*), California lilac (*Ceanothus*), oak (*Quercus*), and gooseberry and currant (*Ribes*).

About 60 million years ago, when the Mediterranean Basin had a moist, subtropical climate, sclerophyll oak and laurel forests were found in the Sahara Desert. As in North America, the climate gradually became drier, and sclerophyll plants that were preadapted to drought were able to spread, survive, and diversify whereas a far greater number of other wet-climate species succumbed.

In addition to these mostly subtropical origins, the mediterranean climate zones of North America and Eurasia also had smaller contributions from a more temperate northern flora.

Southern Hemisphere. As in the northern hemisphere, the shrubs and trees of mediterranean climate areas in the southern hemisphere have predominantly subtropical origins together with some more temperate climate contributions (Axelrod, 1973; White, 1986; Hopper, 1992; Linder et al., 1992; Barlow, 1994; Dodson, 1994; White, 1994; Arroyo et al., 1995; Cowling et al., 1996; Hopper et al., 1996).

Many sclerophyll elements of the Chilean flora probably evolved from ancestors in rainforests that grew close to their present location but in a warmer climate (Arroyo et al., 1995). Among the three mediterranean climate areas of the southern hemisphere, Chile is alone in having dense forests extending far to the south. In the moister, southern part of the mediterranean climate area of Central Chile, flora from a temperate, southern origin is prominent (Rundel, 1981), with southern beech (*Nothofagus*), myrtle (Myrtaceae), *Drimys* (Winteraceae), and conifers of the genus *Podocarpus* being examples.

In South Africa, as in the other mediterranean climate regions, most species have semitropical or tropical origins (Goldblatt, 1978; Linder et al., 1992). The growing diversity and distinctiveness of plants was presumably fostered by the isolation of the Cape Region by a belt of increasing heat and aridity to the north. Africa was situated about 15° south of its present position when flowering plants became established and had a substantial area with a temperate climate.

In Australia, plants with sclerophyll leaves survived on the margins of dense forest as rainfall became scarcer. The myrtle and protea families both contain primitive genera that are still found in rainforest. Many Western Australian plants still show growth peaks in the summer, in accord with their ancestors that originated in climates with summer rainfall. The distinctive flora of the southwestern tip of Australia probably developed during a prolonged period of isolation from eastern Australia, separated first by an inland sea over 100 million years ago in the Cretaceous Period, and later by an enormous bay covering what is now the Nullarbor Plain. Subsequently, this area remained a physical barrier for plant migration by virtue of its arid or semi-arid climate.

ICE AGES AND THE RECENT APPEARANCE OF A MEDITERRANEAN CLIMATE

The last three million years were marked by about 30 ice ages, during which glaciers advanced to scour and cover much of North America and Eurasia (van Andel, 1994). The ice ages had a periodicity of about 100,000 years, with intervening warmer periods lasting roughly 10,000 years. Although the present mediterranean climate areas were not covered by glaciers, their average temperature declined substantially, by 3 to 12° F (2 to 7° C). Nevertheless, some parts of the Mediterranean Basin remained sufficiently mild to allow survival of semi-tropical genera that date back 20 to 50 million years, such as carob (*Ceratonia*), myrtle (*Myrtus*), grape (*Vitis*), oleander (*Nerium*), plane (*Platanus*), and olive (*Olea*) (Polunin and Huxley, 1990). Northern European examples of these plants did not survive the ice ages.

The warmer intervals between ice ages were marked by a retreat of the glaciers and the polar ice caps, resulting in higher sea levels from ice-melt and reducing the land area (Dodson, 1994). It was only during these warmer periods that the mediterranean climate appeared, becoming more profoundly dry in the summer after each successive ice age (Axelrod, 1973) (Fig. 2.03).

In California, Chile, and parts of the Mediterranean Basin, summer drought may have been further increased by the concurrent uplift of major mountain ranges, the Sierra, Andes, and Alps, respectively. Erosion of newly uplifted mountains also created a variety of soil conditions which, together with a changing climate, favored the diversification of plants.

Southern continents, other than Antarctica, were largely free of ice during the ice ages. Chile alone of the three southern hemisphere mediterranean climate areas had substantial glaciation (Villagran, 1995). The southern hemisphere had less land in the cooler latitudes than the northern hemisphere. During the past three million years, Africa and Australia did not reach beyond 45° latitude, and South America projected its narrow tip only as far as about 55°. Furthermore, the Antarctic ice cap was prevented from extending to adjacent continents by the broad Southern Ocean, which surrounded the Antarctic continent.

Although close relatives of present-day sclerophyll species of mediterranean shrubs and trees were living long before a mediterranean climate appeared, many herbaceous (non-woody) perennials and annuals evolved more recently, during the ice ages (Axelrod, 1973 and 1989). The short reproductive cycle of herbaceous plants makes it possible for them to evolve rapidly with changing environmental conditions. This characteristic is an advantage in responding to selective pressures of repeated ice ages. Ice ages resulted in extinction of many species, creating opportunities for survivors to spread and diversify with less competition. The last glacial maximum was about 18,000 years ago, recently enough to affect stone-age man. The development of agriculture about 7,000 years ago and our present-day climate fall into one of the relatively brief, warm intervals.

Northern Hemisphere   **Southern Hemisphere**

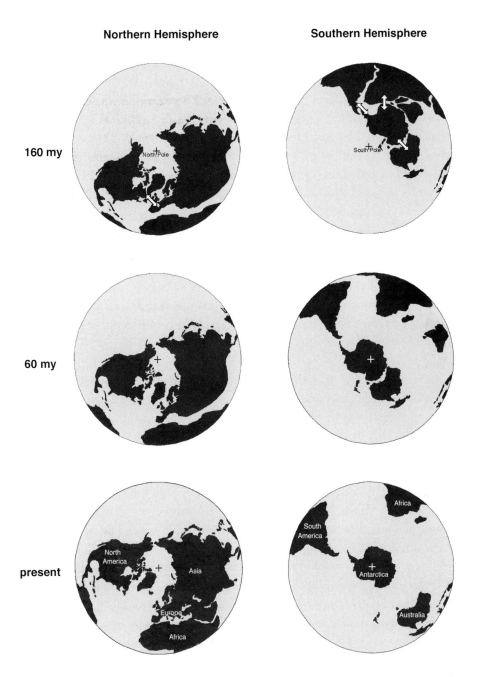

FIG. 2.04. *Plate tectonics: a basis for shared plant families and genera. In the northern hemisphere (the three maps on the left), the proximity of Europe, Greenland, and North America between 160 and 60 million years ago (my) provided a basis for shared flora and a pathway for plant migration shown by the arrow in the top left panel.*

In the southern hemisphere, Africa, Australia, and South America were also much closer, the last two being still connected through a temperate Antarctica 160 my (arrows in the top right panel). Australia and South America remained close to Antarctica 60 my. The lower maps depict the present position of the continents. (Based on van Andel, 1994, and Smith et al., 1981.)

WORLD DISTRIBUTION OF RELATED PLANTS

A Legacy of Continental Breakup

A few plant families, such as protea (Proteaceae), and many genera, such as oak (*Quercus*) and pine (*Pinus*), are found on both sides of the Atlantic and Pacific oceans. This phenomenon puzzled botanists for over a century. For example, Sir Joseph Hooker, surgeon and botanist on the British explorations of the Southern Oceans in 1839 and 1843, was surprised to find forests of southern beech (*Nothofagus*), *Podocarpus,* and *Araucaria* and shrubs of the protea family on southern continents and islands that were widely separated by ocean (Hooker, 1853).

On his return, he discussed his observations with Charles Darwin (Huxley, 1918). They agreed that the continents might previously have been connected through an Antarctica with a milder climate. Their speculation has since gained strong support from the study of plant fossils and of continental movement.

Certain of the close relationships that exist between shrubs and trees of distant regions are now attributed largely to the breakup and movement of continents over the course of time by a process called plate tectonics.

Northern Hemisphere. Plant similarities between California and the Mediterranean Basin are thought to be attributable to the close contact that once existed between the northern parts of Eurasia and North America (Fig. 2.04). These great land masses started to break apart more than 160 million years ago, during the age of dinosaurs and conifers, from a larger common continent. Fossil evidence indicates that forests extended to within 5° of the north pole. The coast redwoods (*Sequoia sempervirens*), now restricted to California and southern Oregon, once grew in Greenland, Europe, and much of Asia. Europe, Greenland, and North America remained in close proximity even 60 million years ago, when flowering plants had become dominant. Oaks and pines were found in much of the northern hemisphere, as they are today.

Distances across the semi-tropical mid-Atlantic Ocean were greater than in the north. Correspondingly, genera of semi-tropical origin from Europe and North America are not as closely related (Axelrod, 1973; Raven, 1973). For example, prominent Mediterranean Basin genera, such as myrtle and olive, and California genera, such as manzanita and California lilac, share no close relatives with the other region, probably because the ocean formed a much broader barrier in the south.

Southern Hemisphere. About 160 million years ago, Africa, Australia, New Zealand, South America, and Antarctica had a mild, moist climate and were part of an enormous southern continent that is referred to as Gondwana (Fig. 2.04). Africa was the first to break off, but Australia and the southern tip of South America remained close to a temperate Antarctica 60 million years ago, early in the Tertiary. Evidence for an earlier, mild Antarctic climate comes from fossil remains on the Antarctic Peninsula, which points toward the southern tip of South America. In this presently cold and barren environment, there are petrified tree trunks of evergreen conifer species of *Podocarpus* and *Araucaria* that show wide and uniform growth rings, indicative of favorable climatic conditions (White, 1994).

The ancient land links of the three continents through Antarctica are thought to account for certain shared families among the southern hemisphere areas with a mediterranean climate. The protea family (Proteaceae) and the rush-like Restionaceae are both prominent elements in the mediterranean climate flora of South Africa and Australia, and are represented in the Chilean flora. Nevertheless, the period during which all three land masses have been separate has been long enough to allow members within the same family to evolve into distinctive genera and species.

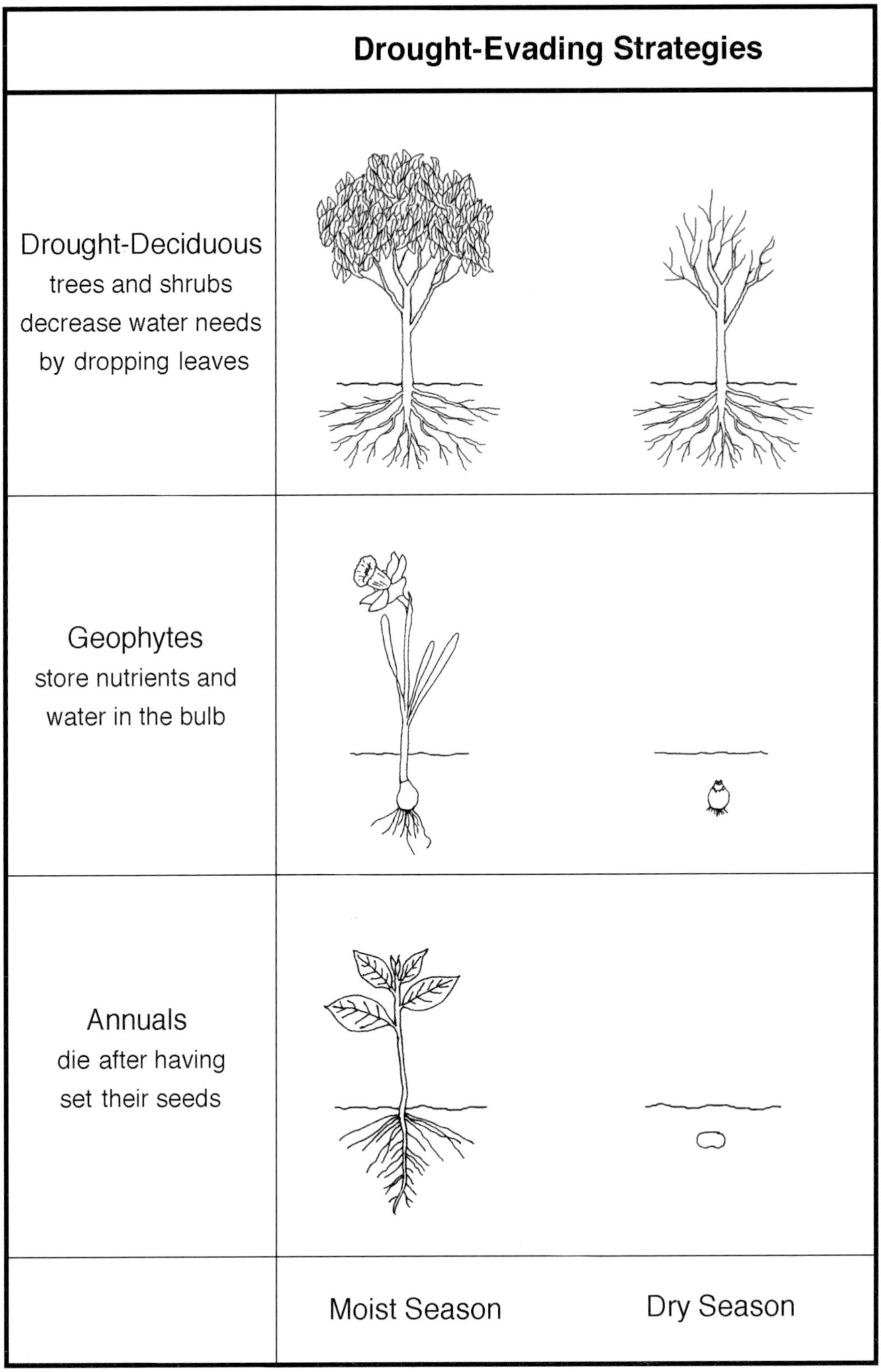

Drought-Evading Strategies

Drought-Deciduous trees and shrubs decrease water needs by dropping leaves

Geophytes store nutrients and water in the bulb

Annuals die after having set their seeds

Moist Season   Dry Season

FIG. 3.01. *Drought-evading adaptations of plants. (Based on a figure in Montenegro et al., 1988.)*

3.

PLANT ADAPTATIONS

Plants that have survived and diversified in regions with a mediterranean climate have done so only by having special characteristics. Most of these are adaptations that favor survival during the period of summer drought and include the following:

Tough, drought-resistant (sclerophyll) leaves are thick and have a coating or cuticle that protects them from dehydration.

Drought-evading foliage. Drought-deciduous plants drop most or all of their leaves during summer drought (Fig. 3.01). They are termed drought-deciduous to distinguish them from winter-deciduous plants. Seasonally dimorphic plants grow two sets of leaves: soft, drought-sensitive ones in the winter and spring, which are replaced by smaller, more drought-resistant ones in the summer.

Bulbs and other geophytes retain their food and water stores underground during the drought season (Fig. 3.01).

Annuals circumvent drought by completing their life cycle before it occurs. The mature seeds are able to survive until the next rainy season or fire triggers their germination.

Roots of many mediterranean shrubs and trees have a dual water-seeking strategy. Seedlings develop their tap roots rapidly and older plants also grow a mat of finer roots close to the soil surface.

Surviving wildfires. Dry heat at the end of the summer and in the early autumn creates conditions for wildfires, which were frequent even before humans had evolved to set them. Mediterranean climate plants are typically fire-adapted.

Nutrient-poor soils are especially widespread in Western Cape Province and Australia, where plant adaptations to these conditions are prominent.

Examples of plants with these characteristics in each of the mediterranean climate areas are shown in Table 3.01.

TOUGH, DROUGHT-RESISTANT FOLIAGE

Evergreen, sclerophyll leaves that are typical of many mediterranean climate plants are tough and leathery and resist dehydration (Fig. 3.02). This leaf structure differs from the larger, softer, and more pliable leaves that are characteristic of the winter-deciduous plants of wet, temperate climate regions of the eastern United States, northern Europe, and East Asia.

In California chaparral, sclerophyll leaves are typically three times as thick as those of

TABLE 3.01. PLANT ADAPTATIONS: EXAMPLES FROM THE FIVE MEDITERRANEAN CLIMATE REGIONS

Adaptation	California	Chile	Cape	Australia	Med. Basin
Sclerophyll foliage	manzanita (*Arctostaphylos* spp.)	litre (*Lithraea caustica*)	*Protea* spp.	*Eucalyptus* spp.	tree heath (*Erica arborea*)
Drought-deciduous	black sage (*Salvia mellifera*)	wild coastal fuchsia (*Fuchsia lycioides*)	*Pteronia nitida*	not prominent	spiny broom (*Calicotome villosa*)
Geophytes	*Brodiaea* spp.	*Alstroemeria* spp.	*Gladiolus* spp.	orchids (Orchidaceae)	*Narcissus* spp.
Annuals	meadow foam (*Limnanthes* spp.)	mariposita (*Schizanthus pinnatus*)	daisy (*Ursinia cakilefolia*)	splendid everlasting (*Helipterum splendidum*)	annual daisy (*Bellis annua*)
Recover from fire by sprouting	scrub oak (*Quercus dumosa*)	litre (*Lithraea caustica*)	*Protea nitida*	mallee (*Eucalyptus* spp.)	scrub oak (*Quercus coccifera*)
Fire-enhanced seed release or germination	most California lilacs (*Ceanothus* spp.)	tevo* (*Trevoa trinervis*)	*Protea* spp.	*Banksia* spp.	*Cistus* spp.

*Fires of natural origin are rare in the Chilean matorral, and fire-stimulated germination of seeds is unusual, in contrast to the other regions (Muñoz and Fuentes, 1989).

deciduous species. Much of this thickness results from a protective, waxy surface cuticle and an abundance of internal supportive tissues. Thick leaves represent a substantial investment of energy by the plant. To a large extent, the energy required to produce such thick leaves is repaid by their relatively long lifespan. Most sclerophyll evergreen leaves last for two years and are dropped primarily during the summer dry period.

Leaves lose water mainly through small surface pores that are lined by specialized guard cells, regulating the size of the opening to a chamber underneath (Fig. 3.03). These humid chambers with their guard cells are called stomata. The guard cells function as hydraulic valves, adjusting pore size according to light intensity, temperature, and humidity. Typically, the openings of stomata of sclerophyll leaves are smaller than those of softer, deciduous leaves, reducing water loss. Stomata of some leaves are sunken in pits or grooves that may be further obstructed by hair or wax tubules. Water loss is also decreased by the position of stomata on the more shaded undersides of sclerophyll leaves.

Although a reduction of water loss protects leaves from drying out, it also slows the rate of photosynthesis by limiting entry of carbon dioxide from the air. Photosynthesis is essential for plant growth and reproduction. It uses the energy of the sun to convert water

Leaf cross sections x 250

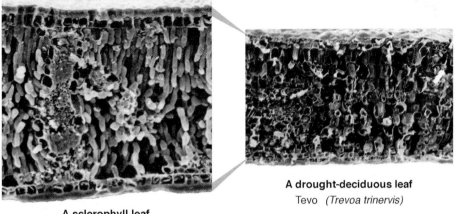

A drought-deciduous leaf
Tevo *(Trevoa trinervis)*

A sclerophyll leaf
Bollen *(Kageneckia oblonga)*

Fɪɢ. 3.02. *Scanning electron micrograph showing cross sections of a sclerophyll leaf on the left and a deciduous leaf on the right. Both are from common plants in the mediterranean climate region of Central Chile. (Reproduced with permission from* Montenegro, 1984.)

Two Stomata

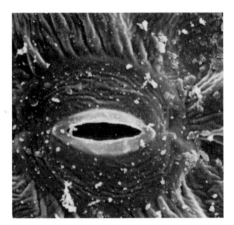

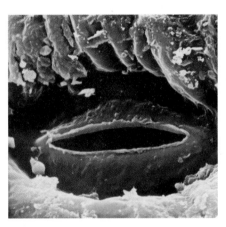

Litre *(Lithraea caustica)*
x 4000

Coliguay *(Colliguaya odorifera)*
x 500

Fɪɢ. 3.03. *Scanning electron micrographs of stomata from the leaves of two sclerophyll plants from Central Chile. At these high magnifications, the stomatal openings look like a mouth. Appropriately enough, this is where leaves exchange gases with the air, taking in carbon dioxide and releasing oxygen and water vapor to the air. (Reproduced with permission from* Montenegro, 1984.)

and carbon dioxide into sugar and oxygen. A decreased rate of photosynthesis is a price paid by sclerophyll plants for their ability to survive summer drought.

Plants with sclerophyll leaves are typically evergreen, an adaptation that allows photosynthesis to continue through most of the year. The rate of photosynthesis slows in winter, owing to cool weather, shorter days, and a lower angle of sunlight. The highest rate of photosynthesis for many evergreen plants is in spring and autumn.

Sclerophyll leaves generally have a high fiber content and low nutritional value, making them less attractive to grazing animals and insects. Many sclerophyll trees and shrubs also produce aromatic oils that repel animal predators.

C

A

D

B

FIG. 3.04. *Vertical, sun-evading leaves are found on many species of* A. *manzanita* (Arctostaphylos) *in California, B.* Leucospermum *in the Cape Region, and C.* Eucalyptus *in Australia.*

D. The gray, hairy leaves of the silver tree (Leucadendron argenteum) *from the Cape Region, E. lavender cotton* (Santolina chamaecyparissus) *from the Mediterranean Basin, and F. beach sagebrush* (Artemisia pycnocephala) *from California protect them from dehydration by reflecting sunlight.*

E

F

DROUGHT-EVADING FOLIAGE

During the drought of summer, plants have to cope with more prolonged and intense sunlight, as the sun rises higher in the sky. Intense, direct sunlight is particularly harmful to plants that have low water reserves. The leaves of some drought-tolerant plants decrease direct exposure to sunlight by permanently orienting themselves in a near vertical position, as is the case with many species of manzanita (*Arctostaphylos*) in California, *Leucospermum* in the Cape Region, and *Eucalyptus* in Australia (Fig. 3.04A-C). Certain species of California lilac (*Ceanothus*) alter their leaf orientation with changes in direction of sunlight. These forms of sunlight avoidance are in marked contrast to the sun-seeking behavior of many tropical and temperate zone plants that have greater access to water.

Some plants, such as the silver tree (*Leucadendron argenteum*), lavender cotton (*Santolina chamaecyparissus*), and beach sagebrush (*Artemisia pycnocephala*) have leaves with a hairy surface or a gray-green color, adaptations that reflect light, decreasing their susceptibility to direct sunlight (Fig. 3.04D-F).

Drought-deciduous plants are able to avoid water loss by dropping their leaves. Many drought-deciduous plants are lower than waist height and grow in drier reaches of the mediterranean climate. They have shallower root systems than sclerophyll shrubs like manzanita, in accord with their decreased water requirement during the months of drought. Examples in the coastal sage scrub of California include the drought-deciduous coastal sage brush (*Artemisia californica*) and purple sage (*Salvia leucophylla*) (Fig. 3.05A).

Plants may also evade drought by seasonal dimorphism, the production of two types of leaves at different seasons (Margaris, 1981). This adaptation is most typical of low shrubs with soft foliage that dominate coastal sage scrub of southern California and garrigue of the Mediterranean Basin. Such shrubs include purple sage (*Salvia leucophylla*) in California

A

FIG. 3.05. A. *Drought-deciduous plants, such as purple sage* (Salvia leucophylla) *in California, have lush, soft foliage when they bloom in early spring, but evade the summer drought by losing most of their leaves. B. Dimorphic leaves are found on Jerusalem sage* (Phlomis fruticosa), *which is native to the eastern half of the Mediterranean Basin. It has a lush appearance with large leaves when it blooms in early spring. Toward summer, new leaves are smaller and more drought-tolerant. Both are in the mint family.*

B

and Jerusalem sage (*Phlomis fruticosa*) in the Mediterranean Basin (Fig. 3.05). The winter and spring foliage of these plants is far softer and less drought-resistant than the evergreen sclerophyll vegetation of chaparral or maquis. These leaves are supplanted in late spring and summer by much smaller leaves that grow on new side shoots. Older, less drought-tolerant leaves then gradually fall off, starting at the bases of their stems, as the summer dry season progresses (Margaris, 1981; Westman, 1982, 1983). Smaller, more drought tolerant leaves lose less than one-fifth of the water lost by the winter-spring foliage.

Many drought-deciduous and seasonally dimorphic plants of California and the Mediterranean Basin are noted for the intense fragrance of their foliage, especially during hot weather. This fragrance results from the release of aromatic oils from leaves during the heat of the day. Although the fragrance and flavor of many types of aromatic leaves are appreciated by humans, they protect the plant by being distasteful to grazing animals. It has also been proposed that the loss of water vapor from these leaves is diminished in hot weather as leaf oils evaporate into the stomatal chamber.

ROOTS WITH TWO WATER-SEEKING STRATEGIES

The pattern of root growth of mediterranean climate shrubs and trees enhances their rate of survival during summer drought. Many species have a dual root system, with a thick, deep taproot and a mat of finer roots closer to the soil surface (Figs. 3.06 and 3.07) (Kummerow, 1981). Young seedlings typically have rapidly growing taproots that penetrate downwards through the soil to keep ahead of the progressive drying that begins at the soil surface. Seedlings of the Chilean tree, quillay or soap-bark tree (*Quillaja saponaria*), can grow a three-foot (1 m) deep taproot in their first year (Canadell and Zedler, 1995). If plants like this survive their first dry season, they add a network of lateral roots that grow

FIG. 3.06. *Root growth. A vertical profile view through a stand of mixed chaparral in Echo Valley, California (inland from San Diego), shows that large shrubs have thick, deep roots, some of which were forced into horizontal growth by bedrock. In addition, there is a mat of finer roots near the soil surface. The scale represents 1 meter. A=Adenostoma fasciculatum; A.p.=Arctostaphylos pungens; C=Ceanothus greggii; H=Haplopappus pinifolius; E=Eriogonum fasciculatum. (From Kummerow, 1981.)*

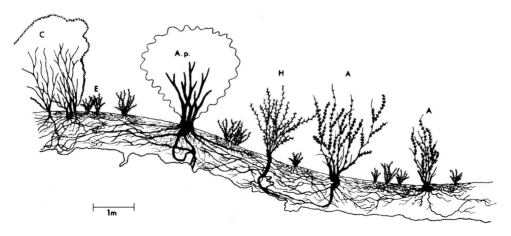

near the surface of the soil where there is better access to nutrients and to moisture from light rains and coastal fog (Kummerow, 1981). An example of an extensive dual system of deep and lateral roots is found among the tall jarrah trees (*Eucalyptus marginata*) of Western Australia, with taproots that reach down as far as 130 feet (40 m). Their lateral roots can extend to the same diameter (Canadell and Zedler, 1995). Among California oaks, the taproots of blue oak, valley oak, and black oak can reach 70 feet (21 m) below the soil surface, where there is likely to be year-round moisture.

Shrub roots penetrate the soil to a more modest degree. In kwongan of Western Australia, for example, only seven of 43 woody species have tap roots that extend more than seven feet (2 m) into the soil. However, the horizontal area of the lateral root systems can be ten to 20 times greater than the plant canopy. Chamise (*Adenostoma fasciculatum*) will extend its lateral roots to as much as seven times its canopy area

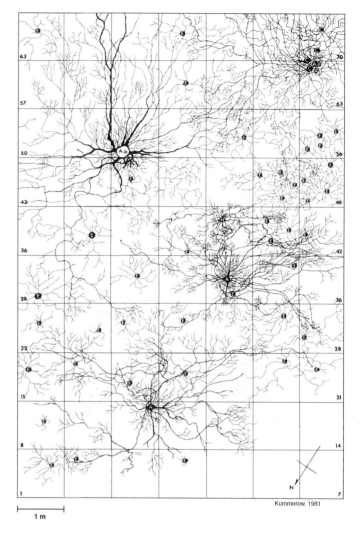

Kummerow, 1981

1 m

FIG. 3.07. *Root growth. A horizontal view of an area of chaparral in Echo Valley, California, shows a wide extension of fine roots, well suited to efficient extraction of moisture near the soil surface. The abbreviations are the same as those used in Fig. 3.06. (From Kummerow, 1981.)*

where soils are shallow in California chaparral. In areas with the most prolonged summer drought, the root mass makes up more than half the weight of woody plants. In contrast, in wetter areas or those with year-round rain, more than half the weight consists of aboveground branches and leaves.

SURVIVAL FOLLOWING WILDFIRE

The risk of wildfires in mediterranean climate areas is much greater than in adjoining semi-arid and rainforest regions. Plants in semi-arid areas accumulate less fuel, and the rainforest typically is too moist. Natural fires are less common in Central Chile than in the other mediterranean climate areas because scrub vegetation is not as dense and lightning is rare (Fuentes et al., 1994).

Sprouters and Seeders. Shrubs are referred to as sprouters or seeders according to which of two strategies they employ to survive a fire (Rundel, 1986; Keeley and Keeley, 1986). Most species follow either one strategy or the other, but not both. Sprouters resprout after a fire from burls or root crowns (also called lignotubers), which are swollen, woody structures at the base of the trunk (Canadell and Zedler, 1995). These structures range in diameter from about an inch (2.5 cm) to more than three feet (1 m).

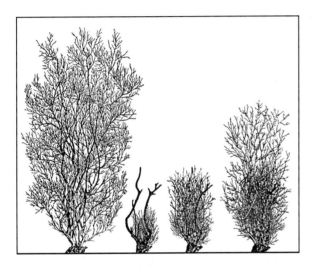

FIG. 3.08. *Recovery by sprouting after a fire. A chamise bush (Adenostoma fasciculatum), and the same plant one, two, and four years after burning. After the first rainy season, new growth begins as resprouting from a burl at the base, but the charred larger stems are still prominent. During the second year, the shrub has doubled its height. Four years after the fire, it is well on its way to the original size. (From Sampson, 1944.)*

In chaparral, at least half of the woody species have the capacity for resprouting from root crowns (Kruger, 1983). Fig. 3.08 shows drawings of a chamise bush (*Adenostoma fasciculatum*) with the same plant one, two, and four years after burning. Fig. 3.09 illustrates the early recovery both by seeding and sprouting after a large October, 1995, forest fire in a bishop pine (*Pinus muricata*) forest. The photographs were taken in the following spring in March, 1996, and a year later in April, 1997, in Point Reyes National Seashore, one hour north of San Francisco. In Australia, a plant community known as mallee is dominated by species of *Eucalyptus* that form dense, tall shrubs with multiple, small upright trunks (Fig. 3.10). After a fire, these plants resprout from unusually large, horizontal, underground lignotubers.

Many seeders in chaparral have unusually long-lived seeds that accumulate in leaf litter and become buried in the top layer of soil. Types of seeds that can survive in the soil for very long periods are referred to as "refractory" or disturbance-dependent (Keeley, 1995). Refractory seeds sprout poorly or not at all when simply watered. Instead, they are stimulated either by heat or by chemicals in the ash of burnt vegetation. After a fire, seeds that have been dormant for as long as decades are released from their dormancy.

Intense brush fires can drive temperatures at the soil surface to well above 1000°F (about 500°C) (Fig. 3.11). But this degree of heat is not sustained for long, since the fire rapidly moves on as surface fuel is consumed. Under a mere two inches (5 cm) of soil

A

B

C

FIG. 3.09. *Recovery after a large October, 1995, forest fire at Point Reyes National Seashore, one hour north of San Francisco. The first three photos, A. to C., were taken in March, 1996. A. shows the charred remains of bishop pines (Pinus muricata), closed-cone pines whose cones retain seeds until they are stimulated to open after a fire. Signs of early recovery include new bishop pine seedlings (B) and coyote brush (Baccharis pilularis) sprouting from the base of its burned branches (C). In April 1997, a lush growth of lupine (Lupinus arboreus) covers the ground (D).*

D

FIG. 3.10. Mallee is the term given to Eucalyptus species that grow as dense, tall shrubs with multiple trunks that emerge from large, horizontal, underground lignotubers. A recent fire in this area of mallee only destroyed the low-growing plants, but had the Eucalyptus also been burned, it could have resprouted from the protected, underground lignotubers.

insulation, the temperature during an intense fire rises to only about 170°F (75°C) (DeBano et al., 1977). Interestingly, many types of seeds are able to survive short periods of high temperature. For example, of 17 herbaceous species found in California chaparral, dry seeds of all species were able to germinate after being placed in a 300°F (150°C) oven for five minutes (Sweeney, 1956). Seeds from five species were even able to withstand 340°F (170°C) for five minutes, but this was close to the limit of temperature tolerance. Moist seeds, as are found in the ground during the rainy season, are much more sensitive to heat. This could help to account for the observation that vegetation following a wet-season fire is different from that seen growing after the more common dry-season fires.

In the Western Cape and Australia, many woody plants retain their seeds in the canopy for many years in hard, woody capsules. Seeds are then released after a fire (page 154, Fig. 8.09). Whether seeds have been stored in the soil or the canopy, they are likely to find favorable conditions for germination after a fire. Ashes from the burned shrubs and leaf litter release nutrients into the soil. In addition, leveling of mature shrubs eliminates competition for light, water, and soil minerals.

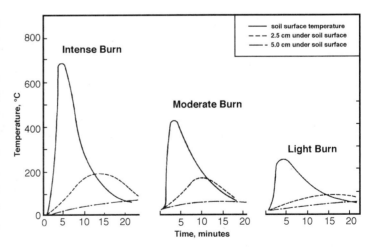

FIG. 3.11. Insulating effect of soil. Fires that create temperatures over 1000°F (about 500°C) at the soil surface of chaparral allow many types of seeds to survive a mere 2 inches (5 cm) under the soil surface, where the temperature does not exceed about 170°F (75°C), a temperature that dry seeds of many species can survive. (From DeBano et al., 1977.)

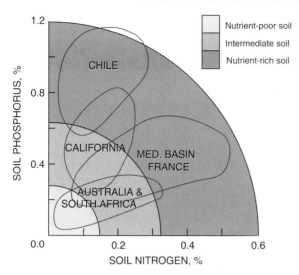

FIG. 3.12. Soil nutrients. The soil in mediterranean climate areas of Western Cape Province, South Africa, and Australia is typically very poor in nitrogen and phosphorus. Most soils in the other three areas are richer in these nutrients. (From DiCastri, 1991.)

Whether shrubs reestablish themselves by sprouting from a burl or root crown or by seeding, it takes many years for them to reach their former size. In the meantime, the space that has been created promotes the luxuriant growth of herbaceous plants, including annuals and bulbs, with the first rainy season.

NUTRIENT-POOR SOILS

Much of the mediterranean climate flora has to survive in unusually nutrient-poor soils. Plant species have developed elaborate mechanisms either to increase the efficiency of nutrient uptake or to tap unusual sources of nutrients (pages 122 and 156). These adaptations are most varied and common in the two regions with the most nutrient-poor soils, Western Cape Province and Australia (Lamont, 1983) (Fig. 3.12).

Root Fungus. A prominent mechanism for enhancing nutrient absorption in many parts of the world is the development of a symbiotic or mutually beneficial relationship between the plant and a fungus that grows on its roots (Lamont, 1983) (Fig. 3.13). A symbiosis of a fungus with the root of a plant is called mycorrhiza from the Greek for fungus root. Mycorrhiza can account for as much as 40 percent of the weight of a root system. The fungus reaches into the soil beyond the roots to extract poorly soluble forms of nutrients such as phosphorus and nitrogen, which rain and soil organisms release into the ground from leaf litter. The fungus can store nutrients and release them to the plant during the seasons when nutrient supply from the soil has diminished. In return, the fungus obtains carbon and sugar

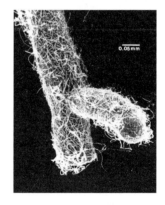

FIG. 3.13. Fungal growth covering the surface of a fine root of a scrub oak (Quercus dumosa), magnified about 100 times by scanning electron microscopy. The Y-shaped root is covered with a hair-like sheath of fungus in what is known as a mycorrhizal association. The fungal threads reach far out into the soil to take up nutrients such as phosphorus that are leached from leaf litter during rains. The fungus subsequently releases the nutrients slowly to the plant when the supply from the soil has diminished. In return, the fungus benefits by receiving carbon and sugar from the host plant. (From Kummerow and Borth, 1986.)

from the host plant. When the fungus dies, nutrients that made up its cells are recycled to the host plant. Mycorrhizal associations are particularly characteristic of pines, oaks, eucalyptus, and the genus *Erica*.

Proteoid Roots. The protea family, Proteaceae, which is prominent in both the Western Cape and Australia, has developed an unusual form of root growth as an adaptation to nutrient-poor soils and periodic drought (Wrigley and Fagg, 1989). These are called proteoid roots and are made up of hundreds of tiny rootlets, resembling cotton (Fig. 3.14). After the first rains of the season, proteoid roots grow in tufts from ordinary roots near the soil surface

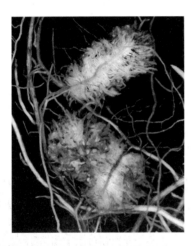

and under accumulated leaf litter. In this location, nutrient levels are highest when organic matter is broken down into a moist mulch by soil organisms. Proteoid roots are short lived and wither after two to three months, leaving only the larger, subsurface roots from which they emerged. Even in the same plant species, proteoid roots are most abundant where the soil is poorest in nutrients. They are very efficient at reaching these scarce nutrients by virtue of their large surface area. Among other plant families that produce similar, temporary root systems are the rush-like Restionaceae and the reeds and sedges of the Cyperaceae (Specht, 1994).

FIG. 3.14. *Proteoid roots are characteristic of the protea family. Soon after the autumn and winter rains accelerate the release of nutrients from leaf litter, these fine, temporary, cotton-like roots grow from the regular, superficial roots of the plant. (Reproduced with permission, Wrigley and Fagg, 1989.)*

Nitrogen-fixing Bacteria. Plants of the pea family have evolved a symbiotic relationship with nitrogen-fixing bacteria that live in swollen nodules on their roots. These bacteria can convert nitrogen from the air into soluble nitrogen fertilizer that can be used by plants, allowing them to thrive in nitrogen-poor soils, where many other species cannot compete. In agriculture, crop rotation takes advantage of nitrogen fixation by clover or alfalfa planted between years of nutrient-depleting crops such as wheat.

Carnivorous Plants. Insect-trapping plants represent a special category of nutrient acquisition. Sundews (*Drosera*), for example, carpet the ground in parts of southwestern Australia (page 157, Fig. 8.11D). *Drosera* have hairy leaves that are topped by transparent drops of sticky secretions that trap insects, which are then digested by the plant.

Plants Parasitic on Other Plants. Southwestern Australia is also the home of the largest plant that is parasitic on other plants, the Western Australian Christmas tree, *Nuytsia floribunda*, which belongs to the mistletoe family (Loranthaceae). The tree was given its name for its brilliant yellow blossoms that are in peak bloom during December. It reaches a height of 40 feet (12 m). The trunk rises from a large lignotuber or underground storage organ for water and nutrients. The root system spreads laterally and attaches to and encircles most other roots that it encounters with collar-like structures called haustoria, which extract water and nutrients from host plants. The tree is considered a hemiparasite because it produces most of its own carbohydrate by photosynthesis.

GOING UNDERGROUND: GEOPHYTES OR BULBS

Geophytes, commonly known as bulbs, have underground food- and water-storage organs that allow plants to resist drought. Geophytes are classified according to the part of the plant that serves as a food storage organ. For example, true bulbs like the onion or lily make highly modified, fleshy leaves that store food and water. Corms, like those produced by *Gladiolus* and *Crocus*, are swollen, vertical portions of underground stem. *Iris* and calla lilies grow from rhizomes, which are horizontal fleshy stems. Aerial portions of tuberous begonias sprout from a tuber, a fat underground stem that is thicker and shorter than a rhizome and does not elongate much as the plant grows.

The underground storage organs of geophytes expend their food and water stores to produce leaves and flowers (Fig. 3.15). They replenish their resources during the wet season from soil nutrients and photosynthesis. During the hot, dry months of late summer, most geophytes become dormant, storing their accumulated food supply while showing little or no sign of life above ground. In autumn or spring, plants sprout again in response to favorable conditions of moisture and temperature. These characteristics are well suited for plant survival in drier regions of the mediterranean climate. Geophytes are common and diverse in the Western Cape and the Mediterranean Basin, especially after fires.

FIG. 3.15. Narcissus *sp., a true bulb. Many members of this genus are native to the Mediterranean Basin.*

A SPEEDY LIFE CYCLE: THE ANNUALS

Annuals are more diverse and abundant in the mediterranean climate than in any other (Raven, 1973). They are concentrated in the semi-arid end of this climate and create spectacular flower displays, especially in open coastal scrub and grassland, after favorable winter and spring rains. Annuals are also found in woodland and chaparral-like vegetation,

Fig. 3.16. *Strawflowers or everlastings of the genera* Helipterum *and* Helichrysum *are in the daisy family and form carpets of flowers in early spring. These* Helipterum *are in Kings Park, Perth, a botanical garden devoted primarily to the native plants of Western Australia.*

particularly as the first in a succession of plants following wildfires. Their short life cycle also makes them well adapted to a brief season of rainfall. Seed from annuals germinates sometime in autumn or winter. Seedlings grow and produce foliage in winter and early spring, when soil is moist and becoming warmer and days are getting longer. With warmer, sunny weather and continued availability of water, the plant concentrates its energy in producing flowers and abundant seeds (Fig. 3.16). After they are dispersed, seeds from many species can survive for long periods, often for many decades, until moisture, light, and temperature conditions are favorable for germination.

Annuals are particularly abundant during a spring that follows a dry-season fire. Fire removes the shade and functioning roots of taller shrubs and trees. Rains following a fire also release nutrients in the ashes of the larger plants. An ability to respond rapidly to increased light and availability of nutrients provides annual weeds a similar advantage in exposed and fertilized soil of gardens and farms.

Annuals make up a varying percentage of the native plant species in the five mediterranean climate regions (Arroyo et al., 1995). In Israel, on the eastern shore of the Mediterranean Sea, about half of the total plant species are annuals. The corresponding figure for California is about 30 percent. Chile has only about 16 percent annuals in its mediterranean flora. The lowest figures of six or seven percent occur in mallee of south-western Australia and fynbos of the Western Cape. Invasive annuals from other mediter-ranean climate areas often compete with the native flora. Weedy annuals originating in the Mediterranean Basin for example have become most readily established in the rich soils of California and Chile.

BENEFITS OF MILD, WET WINTERS

After discussing the adverse circumstances that plants encounter in a mediterranean climate, it is appropriate to emphasize the advantages of a mild, wet winter. The first heavy rains of autumn end the season of drought and greatest fire hazard and begin a period of plant growth. After a few weeks, hillsides turn green. Visitors from colder climates are often surprised by the lush, green appearance of a winter landscape. Most shrubs and many trees retain their leaves and continue photosynthesis throughout much of this season when severe frosts are rare. Early in the new year, there is a peak season of plant growth supported by abundant soil moisture, warmer temperatures, and longer days of sunlight. The wild-flower season begins early and extends through a long spring.

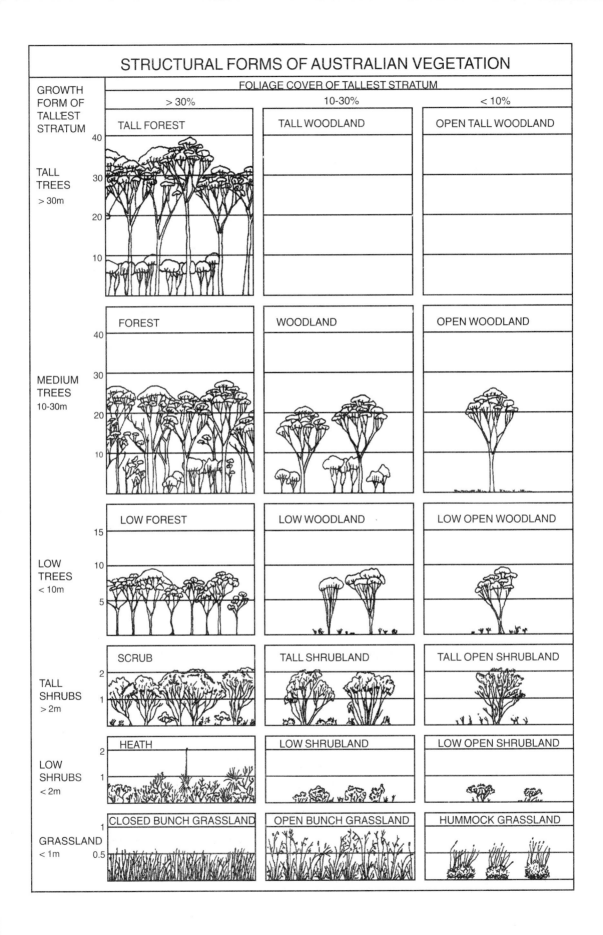

STRUCTURAL FORMS OF AUSTRALIAN VEGETATION

GROWTH FORM OF TALLEST STRATUM	FOLIAGE COVER OF TALLEST STRATUM		
	> 30%	10-30%	< 10%
TALL TREES > 30m	TALL FOREST	TALL WOODLAND	OPEN TALL WOODLAND
MEDIUM TREES 10-30m	FOREST	WOODLAND	OPEN WOODLAND
LOW TREES < 10m	LOW FOREST	LOW WOODLAND	LOW OPEN WOODLAND
TALL SHRUBS > 2m	SCRUB	TALL SHRUBLAND	TALL OPEN SHRUBLAND
LOW SHRUBS < 2m	HEATH	LOW SHRUBLAND	LOW OPEN SHRUBLAND
GRASSLAND < 1m	CLOSED BUNCH GRASSLAND	OPEN BUNCH GRASSLAND	HUMMOCK GRASSLAND

4.

PLANT COMMUNITIES

F*our forms of vegetation are the focus of chapters 5 to 9, which deal with each of the* mediterranean climate regions individually: chaparral-like shrublands; coastal scrub; woodland; and forest.

These groupings are very broad, and each includes many different and distinctive plant communities. Furthermore, these broad forms of vegetation and the names used to describe them differ considerably from one mediterranean climate region to another (Table 4.01).

TABLE 4.01. MAJOR FORMS OF VEGETATION IN MEDITERRANEAN CLIMATE REGIONS

Region	Shrubland	Coastal scrub	Woodland	Forest
California	chaparral	coastal sage scrub	oak	redwood
Chile	matorral	coastal matorral	sclerophyll	*Nothofagus*
Cape	fynbos	strandveld	(scarce)	(scarce)
Australia	kwongan/mallee	kwongan	eucalyptus	eucalyptus
Mediterranean	maquis	garrigue	oak	oak, pine

The terms shrub, woodland, and forest are part of our everyday language. However, when describing plant communities, they are often given more specific meanings. One useful schema to illustrate these terms is based on the mediterranean vegetation of Western Australia (Beard, 1990) (Fig. 4.01). Two terms that can be confusing are shrub and scrub, since they are similar in sound and meaning. Scrub is a vegetation community that is dominated by shrubs, which are perennial woody plants that branch at ground level to form several stems. All of these designations apply to plants with woody stems and branches. Plants without a woody skeleton are called herbaceous. These include bulbs and other perennials as well as annuals. They are primarily low-growing.

FIG. 4.01. *Forest, woodland, and scrub have more specific meanings to botanists than they do to the general public. In this schema, based on Australia, different forms of vegetation are distinguished according to their height and degree of foliage cover. (From Beard, 1990, with some modifications in terminology from AUSLIG, 1990.)*

The major plant communities of the five mediterranean climate areas have similarities of structure and growth patterns. For example, many chaparral-like landscapes could be from any one of the regions when seen from the distance. Closer up, the dominant plant species are distinctive, but nevertheless share many of the features described in the preceding chapter, such as tough, thick, drought-resistant (sclerophyll) leaves. The process by which very different plant species survive a stress like summer drought by similar mechanisms is termed convergent evolution (Mooney, 1977). The degree of this convergence may vary according to the plant communities that are selected for comparison (Hobbs et al., 1995), but there is a broad similarity in respect to an evergreen, sclerophyll habit of growth, spring and autumn growth peaks, and relative dormancy of many species in summer. There are also important differences in vegetation among the mediterranean climate regions (Barbour and Minnich, 1990), particularly in relation to the nutrient richness of the soil, the frequency of wildfires, and the prominence of annuals and geophytes (Chapter 3).

Chapters 5 through 9 describe chaparral-like shrublands, coastal scrub, woodland, and forest in each of the mediterranean climate regions. Shrubland typically occupies drier, hot areas; forest is found in wetter, cooler locations; and woodland is intermediate. The distributions of these plant communities are illustrated by maps in each of the chapters. The vegetation maps depict *natural vegetation* as distinguished from *actual vegetation*. Natural vegetation is the type of plant cover that is believed to have existed before modification by burning, farming, logging, grazing, and urbanization during historic times (Bagnouls and Gaussen, 1957; Lalande, 1968; Küchler, 1977; Bond and Goldblatt, 1984; Beard, 1990). This common convention for depicting natural vegetation can be misleading since it fails to convey the scarcity of native plant cover in places that are intensively farmed and grazed today. The descriptions of plant communities that follow apply to areas of vegetation that have been left relatively undisturbed.

Vegetation maps represent a schematized version of reality. In nature, large expanses of various forms of vegetation may be less common than mosaics of different plant communities. Shrubland, woodland, forest, and coastal scrub often form a complex patchwork, with several forms of vegetation visible from one vantage point.

CHAPARRAL-LIKE SHRUBLANDS

The chaparral-like shrublands of the mediterranean climate regions consist mainly of evergreen shrubs with sclerophyll leaves. In California, chaparral generally consists of a dense, single layer of tall shrubs with relatively few plants in the understory, except after a fire (DiCastri, 1981). In the matorral of Chile, tall shrubs and low trees are more widely spaced, and more herbaceous plants grow in their shelter. Fynbos of the Western Cape and kwongan in Australia are the most diverse in terms of the different heights of plants and number of species. The taller shrublands of the Mediterranean Basin are dense, like chaparral, but may also be variable in height, with tall shrubs and low trees predominating. The chaparral-like vegetation of the Mediterranean Basin will be referred to by its French name, maquis. French terms for the vegetation of the Mediterranean Basin have become widely used, in large part through the long-standing and influential botanical studies at the University of Montpellier

A

B

C

D

E

Fig. 4.02. Chaparral-like shrublands in spring.

A. California: chaparral at Castle Rock State Park in the Coast Ranges with woodland and forest in the background.

B. Central Chile: matorral in the Coast Range near Santiago. Dense shrubs and small trees on the south-facing slope in the middle distance and sparser shrubs on the north-facing slope in the foreground.

C. Western Cape: fynbos vegetation in Paarl Mountain Nature Reserve. The shrubs on the right are in the genus Protea.

D. Western Australia: kwongan with smoke-bush (Conospermum sp.) in the right foreground, a cycad (Macrozamia sp.) in the center, and grass trees (Xanthorrhoea sp.) in the right background.

E. Greece: maquis consisting primarily of evergreen shrubs and low trees of Kermes oak (Quercus coccifera) on the island of Crete.

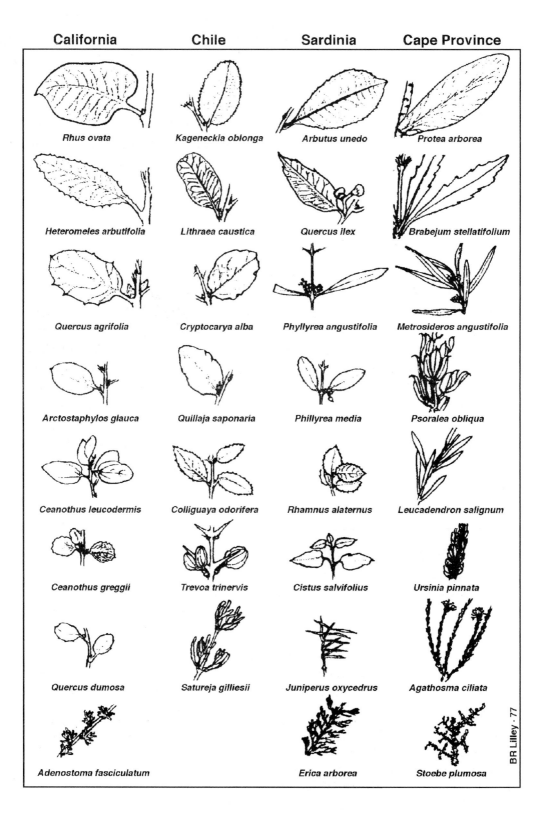

California	Chile	Sardinia	Cape Province
Rhus ovata	*Kageneckia oblonga*	*Arbutus unedo*	*Protea arborea*
Heteromeles arbutifolia	*Lithraea caustica*	*Quercus ilex*	*Brabejum stellatifolium*
Quercus agrifolia	*Cryptocarya alba*	*Phyllyrea angustifolia*	*Metrosideros angustifolia*
Arctostaphylos glauca	*Quillaja saponaria*	*Phillyrea media*	*Psoralea obliqua*
Ceanothus leucodermis	*Colliguaya odorifera*	*Rhamnus alaternus*	*Leucadendron salignum*
Ceanothus greggii	*Trevoa trinervis*	*Cistus salvifolius*	*Ursinia pinnata*
Quercus dumosa	*Satureja gilliesii*	*Juniperus oxycedrus*	*Agathosma ciliata*
Adenostoma fasciculatum		*Erica arborea*	*Stoebe plumosa*

BR Lilley · 77

in southern France. Some of the other terms used in the countries of the Mediterranean Basin include matorral in Spain and macchia in Italy. Seen from a distance, the similarities of some of these landscapes can be striking (Fig. 4.02A-E).

Drought-adapted, sclerophyll leaves of mediterranean climate areas share many features (Fig. 4.03) (Cody and Mooney, 1978). The leaves of common species in four of the regions are arranged by size, with the largest ones at the top. Larger-leaved evergreen species predominate, except in California, where the needle-leaved chamise (*Adenostoma fasciculatum*) is dominant in many areas of chaparral. In the other three regions, plants with the smallest leaves are concentrated in the driest areas. The fynbos of Western Cape Province has the greatest variety in the leaf size of its dominant shrubs, and the sclerophyll leaves are unusually heavy and thick for their surface area. In all of the four areas, as also in Australia, simple leaves predominate, rather than lobed or compound leaves.

COASTAL SCRUB

Coastal scrub is used here as a term to encompass a variety of coastal plant communities that includes mainly low shrubs adapted to wind and salt air off the ocean (Fig. 4.04A-E). Coastal scrub communities in California include northern coastal scrub and coastal sage scrub. Similar plant communities in Chile are referred to as coastal matorral. Somewhat analogous areas are called strandveld or fynbos-thicket mosaic in the Western Cape Province and are particularly rich in geophytes, succulents, and annuals. In Western Australia, sand-heath or kwongan has an unusually large number of species of spring-flowering shrubs. Kwongan includes coastal scrub vegetation, but is also described as being analogous to chaparral, matorral, and fynbos (Pate and Beard, 1984). In the Mediterranean Basin, garrigue is the French term for low, soft-leaved shrublands that include many drought-deciduous species. Other terms used for this form of vegetation are phrygana in Greece, tomillares in Spain, and batha in Israel.

WOODLAND

Woodland consists of trees that are spaced far enough apart to leave room for a variety of shrubs, herbaceous plants, and grasses in the understory. Evergreen and deciduous oaks dominate the woodlands of California and the Mediterranean Basin. *Eucalyptus* is the most prominent genus in Australia. *Nothofagus* woodland is found in Chile. In the

FIG. 4.03. *Leaves of some dominant sclerophyll trees and shrubs of four mediterranean climate areas: California, Chile, Sardinia (Mediterranean Basin), and Western Cape (South Africa). The sclerophyll leaves vary markedly in size from the largest leaves on the top to small, needle-shaped ones at the bottom (the leaves are slightly larger than one-third the natural size, except for the South African Protea arborea and Brabejum stellatifolium, which are a little over one-fourth). The features that they share in common are that all are from plants that remain green in the winter; leaves are thick and tough with a greater weight for their size than deciduous leaves; and the configuration of most leaves is simple rather than compound or lobed. In many cases, similar leaf forms are found across the four areas. (Reproduced with permission, Cody and Mooney, 1978.)*

A

B

C

FIG. 4.04. *Coastal vegetation. Superficial similarities in five mediterranean climate regions. A. Chile: coastal scenery with prominent cacti south of La Serena. B. Western Cape: Cape of Good Hope Nature Reserve, showing the Cape of Good Hope beyond the beach in the distance. C. California: the Big Sur coast of Central California. D. Greece: coastal scenery on the southern coast of Crete. E. Western Australia: Torndirrup National Park near Albany.*

D

E

A

B

C

FIG. 4.05. Woodland. A. California: blue oak (Quercus douglasii) woodland in the Coast Ranges near Santa Barbara. B. South Australia: messmate stringy bark gum woodland (Eucalyptus obliqua) with spring flowers in the understory. C. Greece: woodland in Crete consisting mainly of the deciduous downy oak (Quercus pubescens) just coming into leaf. D. Western Cape: sclerophyll woodland and forest in mountain cleft in the distance with fynbos in the foreground. E. Chile: sclerophyll woodland in Coastal Range near Zapallar.

D

E

mediterranean climate region of the Western Cape, woodland and forest are scarcer than in the other four mediterranean climate areas. There are only scattered woods in sheltered ravines and along stream banks, including trees of the largely southern hemisphere conifer genus *Podocarpus*.

Woodland in mediterranean climate regions is generally found in areas that are suitable for agricultural use, so that much of this vegetation has been converted to pasture or cleared for growing grain. Weeds from the Mediterranean Basin have been introduced into woodlands of all of the other four areas, but they have been most invasive in California and Chile, where exotic grasses thrive on soils that are richer than is typical of the Western Cape or Australia. Fig. 4.05A-E shows examples of woodland in each of the mediterranean climate areas.

FOREST

Forests consist of closely spaced trees that provide a substantial canopy (Fig. 4.06A-E). Notable examples in mediterranean climate areas are the oak and mixed sclerophyll forest of California and the eucalyptus forests of the southwestern tip of Western Australia. Forests require more precipitation than any of the other mediterranean plant communities, but the trees of these and adjacent, wetter areas are nevertheless subject to wildfires and are adapted to survive summer drought and fire. Much of the Mediterranean Basin was once covered by oak forest, but little remains relatively undisturbed. Sclerophyll forest is also severely depleted in Central Chile. The Cape region never had extensive forests. Its largest forested areas benefit from year-round rain along part of the southern coast between Mossel Bay and Port Elizabeth. In the mediterranean climate part of the Western Cape, only remnants of forest remain of the scant amount that existed at the time of early European colonization.

DIVERSITY AND CONSERVATION OF THE MEDITERRANEAN FLORA

High Species Diversity with Many Endemic Plants. Each mediterranean climate area has an unusual degree of plant diversity with a high percent-

FIG. 4.06. *Forest. A. Chile: sclerophyll forest in La Campana National Park near Santiago. B. California: Monterey pine (*Pinus radiata*) forest at Point Lobos State Park at the northern end of the Big Sur coast. C. Western Cape: sclerophyll forest in Harold Porter Botanic Gardens near Hermanus. D. Greece: sclerophyll forest in one of Crete's many gorges. E. Western Australia: karri (*Eucalyptus diversicolor*) forest near Albany.*

A

B

C

E

D

age of endemics (plants that grow nowhere else). Only tropical rainforests of the western hemisphere and southeast Asia appear to have a greater density of plant species than the mediterranean climate regions (Mooney, 1988; WCMC, 1992). The density or diversity of species is substantially less in other temperate areas and still lower in cold winter climates.

The number of plant species and their density in three mediterranean climate areas is compared with larger regions: the entire world, Europe (to the Ural Mountains), Australia, and the United States in Table 4.02. The large land masses of Europe, Australia, and the United States have the greatest absolute numbers of species. But when the number of plant species is expressed on the basis of area, the regions of mediterranean climate vegetation have a more than fourfold greater density of plant species. The mediterranean climate area of California has more plant species than all of central and northeastern United States and adjacent Canada combined, a region that is ten times larger. The Cape region of South Africa and southwestern Australia have even more plant species than California, and they are concentrated within smaller areas of land. The Cape region has the greatest density of plant species by far, due to its small area.

Small areas with a great diversity of plant species are also likely to have the greatest number that are rare and endangered (Cody, 1986; Cowling et al., 1996). Compared with about ten percent rare and endangered species in the large areas listed at the top of Table 4.02, the three mediterranean climate areas have more than twice as many.

The unusual diversity of plant species in mediterranean climate areas is thought to be partly related to their geography and to the changes in climate during geologic time (DiCastri, 1981; White, 1988). Most mediterranean climate regions are sandwiched between moister and drier areas, allowing incursions from adjacent plant communities. Furthermore, this transitional position has made it possible for surviving plants to diversify during climate changes imposed by the advance and retreat of cold temperatures during recurrent ice ages and by continental movement (Chapter 2).

The total number of plant species is shown with percentage of endemic species in Table 4.03. The entire Mediterranean Basin has about 25,000 plant species (Greuter, 1991; Blondel and Aronson, 1995). Italy, Spain, Greece, and France each has about 4,500 or more (WCMC, 1992). By contrast, Northern European countries such as Sweden and the United Kingdom have scarcely more than 1,500 plant species each. Non-Mediterranean Europe as a whole has about 6,000 species in a land area four times as large as the Mediterranean Basin (Blondel and Aronson, 1995).

Central Chile has 2,400 plant species in a much smaller area, the smallest number of the five mediterranean climate regions (Arroyo et al., 1995) (Tables 4.02 and 4.03). Intermediate are California with 4,400 species and the Cape region and southwestern Australia, with 8,550 and 8,000 species, respectively.

The distinctiveness of the plants in each of the mediterranean climate areas can be compared by noting the percentage of species that are endemic. The high percentage of endemic plants in each of the five mediterranean climate areas is noteworthy. In California, Chile, and the Mediterranean Basin, about 50 percent of plant species are endemic (Polunin and Huxley, 1990). Even higher percentages of 68 and 75 percent are found in

TABLE 4.02. PLANT DIVERSITY AND RARITY

Region	Area, km² millions	Plant species, thousands	Species Density*	Percent Rare or Endangered
World	148	250	1.7	10
Europe	5.7	14	2.5	11
Australia	7.6	22	2.9	10
USA	9.4	20	2.1	8
California	0.41	5.1	12	23
Cape Region	0.09	8.6	94	27
SW Australia	0.31	8.0	26	24

*thousands of plant species per million km²

(From Cody, 1986; Cowling et al., 1996.)

PLANT NOMENCLATURE

The system of identifying plants by their genus and species is called **binomial nomenclature**. A **species** represents a group of plants that interbreed among themselves, producing fertile offspring similar to themselves. The **genus** is a group of related species, and the **family** comprises a group of related genera. When plants are identified by their genus and species, these are printed in italics with the genus capitalized. The advantage of binomial nomenclature is that it is used and understood internationally. Common names may be more familiar, but they often refer to more than one species, or a single species can have more than one name. Furthermore, many plants are too unfamiliar to have been given a common name, whereas essentially all known plants have been given a binomial designation.

The relationship between binomial nomenclature and common names is illustrated in the following examples: True oaks belong to the genus *Quercus*, which along with several other genera is in the larger family of Fagaceae. The black oak of California is *Quercus kelloggii*, named after Dr. Albert Kellogg, California's first resident botanist. None of California's 18 species of oaks is found in any of the other mediterranean climate areas. However, the genus *Quercus* is well represented in the Mediterranean Basin by many species that closely resemble their California relatives. Along with oaks, the family Fagaceae also includes beeches (*Fagus*) and chestnuts (*Castanea*), which are close relatives. While the genus *Quercus* is not found in the southern hemisphere, a large group of Australian trees and shrubs came to be known as she-oaks, because the wood resembled oak. Use of the scientific nomenclature makes it unambiguous that these **are unrelated** to true oaks, but belong to the genus *Casuarina* and the family Casuarinaceae.

TABLE 4.03. TOTAL NUMBER OF PLANT SPECIES AND PERCENT OF
ENDEMICS IN EACH MEDITERRANEAN CLIMATE AREA

Location	Species, No.	Endemics, Percent	References
California	4400	48	Raven, 1988
Chile	2400	50	Arroyo et al., 1995; IUCN, 1986
Cape Region	8550	68	Huntley, 1989; Bond & Goldblatt, 1984
SW Australia	8000	75	Hopper, 1992
Mediterranean	25000	50	Greuter, 1991; Polunin & Huxley, 1965

the Cape region and southwestern Australia, respectively (Huntley, 1989; Hopper, 1992) (Table 4.03). Only remote island groups have a greater percentage of endemics, such as Hawaii with 89 percent (Wagner et al., 1990). The extraordinary abundance of plants unique to each of these areas can be put into perspective by a comparison with the United Kingdom, in which only 1.2 percent of plant species are endemic in an area that is similar to the mediterranean climate region of California.

Importance of Conservation. The high percentage of rare and endangered plants in the mediterranean climate areas underscores the importance of conservation, not only of individual species, but also of the relatively undisturbed plant habitats that are disappearing rapidly. Woodland and forest, in particular, have a complexity and characteristic appearance in their virgin or old-growth state that cannot be duplicated by replanting.

The conservation movement has been most successful in the United States, Australia, and South Africa, favored by relatively large acquisitions of public land and strong public support for land conservation. In the Mediterranean Basin a long history of intensive land use and population pressures have combined to diminish areas of relatively undisturbed vegetation to a greater extent. The mediterranean climate area of Chile has become similarly depleted of native vegetation, but over a shorter period of about 450 years.

Protecting substantial areas of mediterranean vegetation is most difficult where the land is best suited for agriculture. Even areas with nutrient-poor soils have been developed at a rapid rate. Early in this century, it was found that adding trace minerals such as copper could make infertile soils of the Western Cape and Australia agriculturally productive. As a result, what is now known as the wheatbelt of Western Australia, much of the mallee region in South Australia, and the Swartland of the Cape Province have been cleared for farming with substantial loss of their native vegetation.

Nearly all undeveloped areas are under increased pressure owing to rapid increases in population particularly over the last century. The vegetation of the mediterranean regions has been especially threatened by development because of a favorable climate, agricultural productiveness, and coastal location, which make the land attractive for urban development, farming, and recreation. Areas of coastal vegetation, chaparral, woodland, and forest have been cleared for housing, crops, grazing, and fuel. The economic desirability of these lands make their acquisition costly when they are not already government-owned. For example, rapid urbanization in Southern California has left few sizable areas of coastal sage scrub, most of which is on private lands that can be acquired for conservation only at very high cost (O'Leary, 1995). Most national parks, forests, and wilderness areas have been established in rugged mountain areas that are unsuited for agriculture and that have limited potential for urban settlement. It is not surprising that the percentage of endangered plant species in the mediterranean climate areas is now second only to that in tropical rain forests.

There is a growing appreciation for the scenic, recreational, and conservation values especially of coastal land, but also of areas such as the oak woodlands of California (Pavlik et al., 1991). Local and international support for preserving representative plant habitats independent of their recreational attractions is of very recent origin. There is a sense of urgency with the realization that native forests, woodlands, and scrub are rapidly disappearing and that numerous species have been irretrievably lost.

It is especially difficult to preserve plant habitat when environmental policy conflicts with strong economic interests. If native plant habitat is on private lands or if logging jobs are threatened, political obstacles increase and the cost of establishing reserves becomes steep and sometimes prohibitive.

Despite all the difficulties in preserving plant habitat during a period of explosive population growth, there is growing public support for identifying and protecting plant and animal communities that represent complete ecosystems. In the 1970s, the concept of Biosphere Reserves was promoted by UNESCO, stimulating worldwide surveys of existing ecosystems to determine which were of highest priority and how they could be adequately protected. This was followed in many countries by legislative action supporting new nature reserves and growing public support for plant conservation. Significant progress is also being made by private organizations such as The Nature Conservancy and various land trusts in California and by organizations with similar goals elsewhere. Although such private land acquisitions are not large compared with national parks and forests, they are disproportionately important because of the focus on the preservation of plant communities that are endangered or have unusual scenic value. Preservation of habitat is increasingly seen as having scientific and educational value as well as representing an ethical and moral responsibility to provide an adequate legacy for the future.

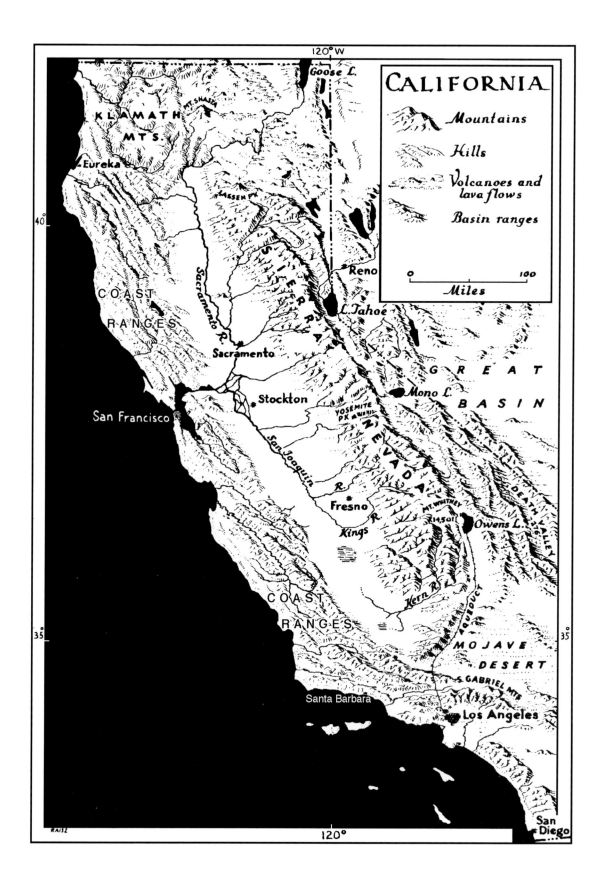

CALIFORNIA

Mountains

Hills

Volcanoes and
lava flows

Basin ranges

0   100
Miles

Goose L.

KLAMATH
MTS.

MT. SHASTA

Eureka

LASSEN PK.

40°

COAST

RANGES

Sacramento R.

Reno

L. Tahoe

Sacramento

San Francisco

Stockton

GREAT

BASIN

Mono L.

S
I
E
R
R
A

N
E
V
A
D
A

San Joaquin

R.

YOSEMITE
PK. 4WHI.

DEATH VALLEY

Fresno

R.

MT. WHITNEY
14,501

Owens L.

Kings

COAST

RANGES

Kern R.

AQUEDUCT

MOJAVE

DESERT

35°   35°

S. GABRIEL MTS.

Santa Barbara

Los Angeles

RAISZ

120°

San
Diego

5.

CALIFORNIA

LANDSCAPE AND CLIMATE

The large part of California that has a mediterranean climate lies west of the Sierra Nevada and its other high mountain ranges (Fig. 5.01). It also includes a portion of coastal Baja California to the south.

The state of California measures about 700 miles (1,125 km) from north to south. Most of the Pacific coastline north of San Francisco is flanked by the Coast Ranges, which rise to no more than about 4,300 feet (1,300 m). South of the Golden Gate, the Coast Range barrier becomes particularly rugged where the Santa Lucia Range drops abruptly into the Pacific along the Big Sur coast. Near Santa Barbara, the Transverse Ranges run eastward, with their highest elevations in the San Gabriel and San Bernardino Mountains of the Los Angeles area. The peaks of the Transverse Ranges reach 6,000 to 11,500 feet (1,800 to 3,500 m) and are often snow covered in the winter. Further south, toward San Diego, the Peninsular Ranges form a lower barrier between desert to the east and the woodland and chaparral vegetation to the west.

Part of the California coastline is densely settled and includes the three largest metropolitan areas of the state, the Los Angeles basin, the San Francisco Bay area, and the San Diego region. However, there are also stretches of coast that are among the least inhabited, most scenic, and least developed of any coastline with a mediterranean climate. Examples are the Big Sur coast in Central California and the so-called empty coast, the roadless area south of Eureka in the northern part of the state. These rugged areas are particularly attractive to naturalists because they retain much of their original vegetation.

California's great Central Valley is the next major topographic feature as one moves inland from the Coast Ranges. This broad, 400-mile (640 km)-long valley once included large freshwater marshes and broad expanses of native bunch grasses. The vegetation of the Central Valley has become more extensively modified than almost any other in the state.

FIG. 5.01. *California topography. Areas with a mediterranean climate extend from the coast to the higher elevations of the Klamath Mountains, the Sierra Nevada, and San Gabriel Mountains. Coast Ranges run parallel to the coast, the broad, flat Central Valley, and the higher mountains to the east. (Adapted from a map by the noted cartographer, Erwin Raisz (1893-1968) with permission of John Wiley & Sons. The original appeared in A Geography of Man, 3rd edition, by Preston E. James, 1965.)*

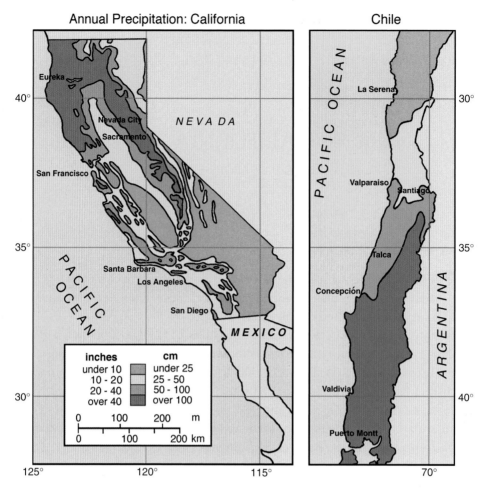

Annual Precipitation: California

Chile

inches | cm
under 10 | under 25
10 - 20 | 25 - 50
20 - 40 | 50 - 100
over 40 | over 100

0   100   200   m
0   100   200 km

FIG. 5.02. *California and Chile: annual precipitation. In California, rainfall is heaviest on the west side of mountain ranges, with an overall decrease in precipitation from north to south. In Chile, the major pattern is the decrease in rainfall from south to north. (Based on Donley et al., 1979, for California and Hoffmann, 1975, for Chile.)*

Rich soil and availability of irrigation water from Sierra Nevada snowmelt have made this one of the world's most productive agricultural areas.

To the east of the Central Valley, foothills rise to the volcanic Cascade Range in the north and to the Sierra Nevada in the central part of the state. Oak woodlands and chaparral vegetation cover the lower elevations of the foothills, which become densely forested with increases in elevation. Parts of the forest are protected by national parks and national forests.

Rainfall

The wettest parts of California's mediterranean climate region are in the north and the driest are to the south (Fig. 5.02). Of the cities shown in Figure 1.09, Eureka on the north coast has the greatest amount of rainfall. Toward the south, San Francisco, Los Angeles, and San Diego become progressively drier. The areas with the heaviest winter rains also have the shortest summer dry seasons, Eureka having fewer than four dry months compared to San Diego with more than eight dry months.

The climate maps of California are shown together with those of Central Chile to illustrate the many similarities between the two. In southern hemisphere Chile, north-south relationships are reversed. Cooler, moister climate is to the south and drier climate to the north. In both places, deserts lie toward the equator and rainforests in the polar direction. More than 85 percent of the rainfall is concentrated in the winter half of the year in the greater part of both regions (Thrower and Bradbury, 1973).

Superimposed on California's north-south moisture gradient are the east-west effects of its mountain chains and its coastal fog. The forested Coast Ranges and western slopes of the Sierra Nevada are rainier than the relatively dry Central Valley in between, creating patterns of rainfall that change markedly from east to west. Precipitation increases by about seven inches with each 1,000 foot increase in elevation up to 8,000 feet (18 cm/300 m) (Barbour et al., 1993). Summer fog is common along the entire Pacific coast, reducing drought stress on coastal vegetation.

Drier areas of California have less predictable rainfall than wetter areas. Several consecutive dry years result in severe water shortages. Recurrent

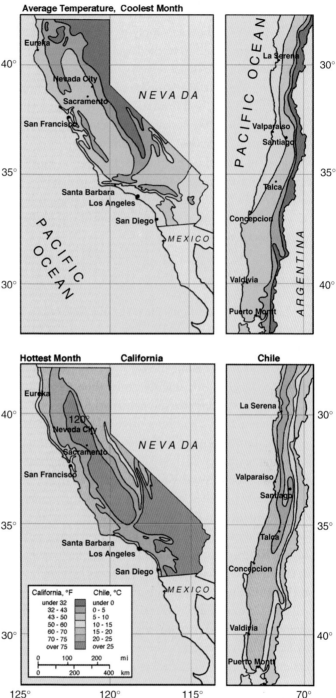

FIG. 5.03. *California and Chile: average temperature of the hottest and coolest months. Central Chile is milder than California, with no area as hot as the California Central Valley in the summer. Because Chile is so narrow, the moderating effect of the Pacific Ocean extends all the way across to the Andes in the east. (Based on Donley et al., 1979; Hoffmann, 1975.)*

cycles of drought increase competition between rapidly growing urban regions and agricultural areas for the limited supply of water. Water supply for large portions of California depends on snowmelt from the high mountain chains that form the eastern boundary of the mediterranean climate region.

During winter, there is a pattern of relatively infrequent but severe storms that cause flooding and severe soil erosion. Areas that have been largely cleared of vegetation are particularly vulnerable because their lack of roots and foliage cover makes them less able to retain soil and absorb water from downpours.

The San Francisco Bay region, the second largest metropolitan area in California, is located near the juncture of moister areas with more than 40 inches (100 cm) to the north and drier ones to the south and east where there is less than 20 inches (50 cm) of rainfall per year. The moister coastal region north of San Francisco supports the largest areas of coast redwood forest.

Temperature

The temperature along the coast of California is mild in the summer and almost frost free in the winter (Fig. 5.03). The difference between the average temperatures of the warmest and coolest months is small, well below 18°F (10°C). North-south differences in average temperature along the coast are also relatively modest, with the Mendocino coast in the north being only slightly cooler than the central Big Sur coast.

Summer temperatures increase markedly as one travels inland. Sheltered valleys that are little more than a half hour drive inland from San Francisco (Walnut Creek and Concord) and Los Angeles (San Fernando and San Bernardino valleys) have average summer highs that are about 25°F (14°C) greater than those along the coast. Residents of these areas and of the Central Valley are prepared to put on extra layers of clothing when they make a day trip to the Pacific coast during the summer, particularly if there is fog.

Summer heat in the Central Valley results from a barrier created by the Coast Ranges. The major break in this barrier is at San Francisco Bay, where the Sacramento and San Joaquin Rivers of the Central Valley drain into the Pacific Ocean. In the summer, cooling fog enters the Golden Gate at San Francisco, moderating the temperatures of the Bay Area and the prime wine-growing regions of Napa and Sonoma counties.

The Central Valley is largely beyond the influence of coastal fog and is hot in the summer. But toward the east, temperatures decrease again with increasing elevation from the foothills to the crest of the Sierra Nevada. The daily and seasonal ranges of temperature are greatest at higher elevations, which have an alpine climate. Both elevation and distance from the coast have important influences on the climate and vegetation (Ornduff, 1974).

Plant Communities

There are many distinctive plant communities in the mediterranean climate area of California (Munz, 1968; Ornduff, 1974; Bakker, 1984; Barbour and Major, 1988; Henson

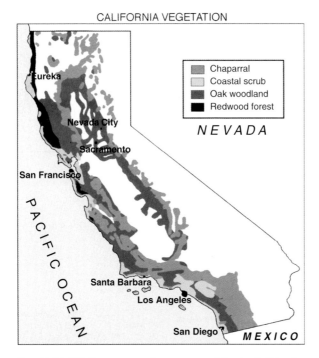

CALIFORNIA VEGETATION

Chaparral
Coastal scrub
Oak woodland
Redwood forest

Fig. 5.04. Mediterranean plant communities in California: chaparral, coastal scrub, oak woodland, and redwood forest. (Based on Küchler, 1977; Hanes, 1988; and Griffin, 1988.)

and Usner, 1993). In Fig. 5.04, these are grouped under broad categories of chaparral, coastal scrub communities, oak woodland, and redwood forest. While species composition of vegetation changes from north to south, the greatest variations over a short distance occur from west to east. This is illustrated by schematic cross sections across the northern, central, and southern parts of the state (Fig. 5.05).

The greatest diversity of plant communities occurs near the coast, as one can appreciate when walking, bicycling, or driving. As one travels inland, striking changes in native vegetation and in the agricultural uses of the land occur from ridge to ridge and mile to mile. Near Bodega Harbor on the Sonoma County coast north of San Francisco, for example, cows and sheep are pastured within the first few miles of the coast, where the seasonal differences in temperature are small, extending the fall to spring season of grazing. A belt of coast redwoods is centered about five miles (8 km) inland, often within a stone's throw of drier oak woodlands, chaparral, and grasslands. One of best apple-growing areas in the country is just beyond, about eight miles (13 km) from the coast. The famous wine regions of Sonoma County begin only with the warmer climate 10 miles (16 km) inland. Although residents of coastal California tend to take this diversity for granted, there are few places in the world that have as great a variety of climate and plant life within such short distances.

The complex mosaic of California's coastal landscape, shown schematically in Fig. 5.06 (Bakker, 1984), could be based on Mount Tamalpais, just north of San Francisco. The figure illustrates how a landscape that might hypothetically start out as chaparral and coastal scrub is modified by factors such as wind exposure, orientation to the sun, character of the soil, and periodic fires.

Chaparral

Chaparral is the most widespread form of vegetation in the mediterranean climate part of California, covering about five percent of the state (Hayes, 1977; Rundel, 1986) (Fig. 5.04) or 8.6 million acres (3.5 million hectares). From a distance, chaparral has the decep-

Northern California (N):
TRINIDAD HEAD-TRINITY ALPS-MODOC PLATEAU

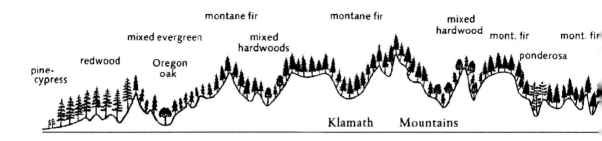

montane fir   montane fir   mixed hardwood

mixed evergreen   mixed hardwoods   mont. fir   mont. fir

redwood   Oregon oak   ponderosa

pine-cypress

Klamath   Mountains

Central California (C):
PAJARO VALLEY-GABILAN RANGE-SAN JOAQUIN VALLEY-
SIERRA NEVADA-OWENS VALLEY-WHITE MOUNTAINS

coastal prairie   mixed hard-wood

blue oak-digger pine

seashore   valley oak savanna   prairie   saltbush   tule marsh   Calif. prairie

Southern California (S):
SAN PEDRO BAY-SAN BERNARDINO MOUNTAINS-
OLD WOMAN MOUNTAINS-COLORADO RIVER

subalpine fir

jeffrey pine   jeffrey pine

pinyon-juniper

chaparral   creosote bush

coastal sagebrush   southern oak   coastal sagebrush   chaparral   coastal sage

seashore

San Bernardino Mtns.

FIG. 5.05. Plant profiles from west to east through Northern, Central, and Southern California show the strong influences of distance from the coast, elevation, and diminishing precipitation from north to south. The orientation map at the lower right shows the location of the three profiles or transects.

The northern transect is near the northern extreme of California's mediterranean climate region. It begins in the west with coastal pine and cypress followed by redwoods. Further inland in the Klamath Mountains are Oregon oaks, mixed evergreens and hardwoods, and montane fir. Toward the east, beyond the mediterranean climate region, is drier Modoc Plateau with sagebrush, jeffrey pine, and juniper savanna.

The central transect starts with coastal scrub and mixed hardwood. There is a belt of valley oak savanna in the upper Santa Clara valley followed by blue oak and digger pine in the Coast Ranges. There is a broad stretch of Central Valley, which might be considered too hot and dry to be truly mediterranean.

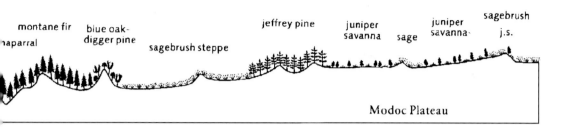

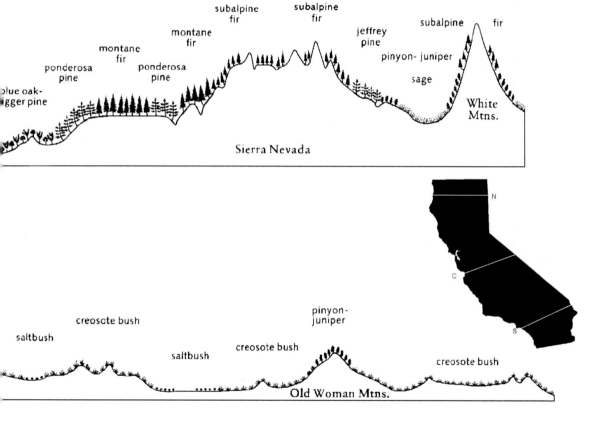

This includes prairie, saltbush, and tule marsh. In the Sierra foothills to the east are the progressively moister and milder zones of blue oak and digger pine, ponderosa pine, and montane fir. Both the Coast Ranges and the Sierra foothills have areas of chaparral which are not shown. The higher part of the Sierra Nevada and the area to the east are too cold to be considered mediterranean in climate.

The southernmost transect first traverses an area with a mediterranean climate, including coastal sagebrush, oak, and chaparral. The colder jeffrey pine and subalpine fir belts are at the higher altitudes in the San Bernardino Mountains. Beyond to the east is primarily desert vegetation with creosote bush and saltbush.

(Based on the map "Natural Vegetation of California" by AW Küchler, from Terrestrial Vegetation of California, 1977, adapted with permission, Donley et al., 1979.)

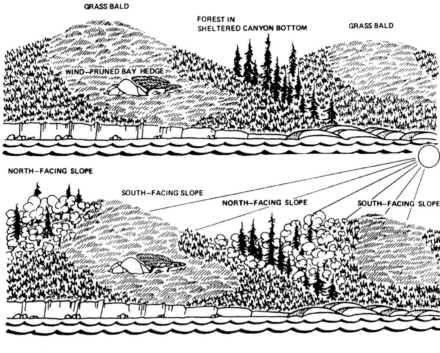

GRASS BALD

FOREST IN
SHELTERED CANYON BOTTOM

GRASS BALD

WIND-PRUNED BAY HEDGE

NORTH-FACING SLOPE

SOUTH-FACING SLOPE

NORTH-FACING SLOPE

SOUTH-FACING SLOPE

OUTCROP OF
INFERTILE SANDSTONE

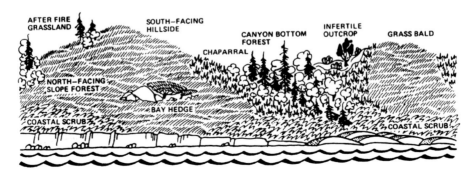

AFTER FIRE
GRASSLAND

SOUTH-FACING
HILLSIDE

CANYON BOTTOM
FOREST

INFERTILE
OUTCROP

GRASS BALD

CHAPARRAL

NORTH-FACING
SLOPE FOREST

BAY HEDGE

COASTAL SCRUB

COASTAL SCRUB

tively smooth look of a low and finely textured plant cover (Fig. 5.07A). However, when one gets close, its height and impenetrability become apparent (Fig. 5.07B). Chaparral consists mainly of tough, woody, evergreen shrubs with dense foliage growing at least to the average height of an adult. Most chaparral species have broad, sclerophyll leaves, with the exception of the common needle-leaved chamise (*Adenostoma fasciculatum*) (Fig. 5.08). Some chaparral shrubs, such as flannel bush (*Fremontodendron californicum*), western red-bud (*Cercis occidentalis*), and California lilac (*Ceanothus*) have showy flowers in mid-spring (Fig. 5.10A, C, E). Late spring and early summer bring red buckwheat (*Eriogonum latifolium*) and farewell to spring (*Clarkia rubicunda*) (Fig. 5.10B, D), growing mainly at the edges of the chaparral, along paths, and in clearings.

The word chaparral originates from the Spanish *chaparro*, meaning place of scrub oak, which in Spain forms a similarly dense and virtually impenetrable form of vegetation. The chaps worn by cowboys to protect their legs from brush derive their name from the word chaparro. To this day, chaparral is best seen from trails that have been cut through it.

In Northern and Central California, chaparral covers much of the Coast Ranges (Fig. 5.04). It is also prominent but more scattered on the western foothills of the Sierra Nevada, including historic sites from the 1849 gold rush. Southern California has the greatest expanse of chaparral, at elevations of 1,000 to as high as 10,000 feet (300 to 3,000 m) in the Transverse and Peninsular ranges (Hanes, 1988). At lower elevations, toward the Pacific coast, chaparral impinges on the metropolitan areas of Los Angeles and San Diego and is bordered by coastal sage scrub.

In the mediterranean climate part of California, chaparral is found where annual rain-

FIG. 5.06. *Patterns on the hills. This series of drawings illustrates some of the reasons for a varied mosaic of plant communities on the coastal hills north of San Francisco.*

The top panel illustrates the relatively uniform vegetation of low coastal scrub and chaparral that one might anticipate within such a small area.

The next panel is a more realistic picture representing the microclimates due to salt-laden wind, fog, and varying soil moisture. Grassland or coastal prairie is found on the wind-blown bluffs and headlands. Some sheltered hollows protect the rounded, wind-pruned shapes of California bay trees and evergreen oaks. Canyon bottoms collect sufficient water and provide shelter for coast redwood forest.

The middle panel illustrates the consequences of northern versus southern exposure. The north-facing slopes are more protected from direct sunlight and provide a cooler, moister environment that supports evergreen oaks and coast redwoods. The hotter, drier, south-facing slopes support chamise chaparral, California sagebrush and black sage coastal scrub, and grasslands that are full of wildflowers in the spring.

The fourth panel depicts the additional influence of the soil. Sandy soils hold less water than clay soils. Serpentine soil high in magnesium supports only certain types of plants. Infertile sandstone modifies the vegetation, as illustrated.

The additional effect of fire is represented in the bottom panel. Summer and early fall wildfires occur every few years and destroy chaparral, woodland, and small forest groves, temporarily favoring the spread of grassland. The resulting vegetation is not only highly varied due to weather, sun exposure, and soil, but changes from year to year according to the duration since the most recent fire.

(From Bakker, 1984, reproduced with permission of the University of California Press.)

fall is between about 12 and 25 inches (30 and 60 cm), roughly corresponding to the parts of the rainfall map that are shown in yellow (Fig. 5.02). Where there is less rainfall, sparser, semi-desert or desert vegetation is found. Where rainfall is more abundant, the growth of woodlands and forests becomes possible.

A

B

Types of Chaparral

Chamise Chaparral. Chamise (*Adenostoma fasciculatum*) belongs to the rose family (Rosaceae) and has the widest distribution of any major species in the chaparral (Hanes, 1977; Henson and Usner, 1993). It exists in almost pure stands, referred to as chamise chaparral or *chamisal* in Spanish. Chamise tolerates dry, nutrient-poor soil and especially favors hot, south- and west-facing slopes. The shrubs have wiry branches and bundles of drought-resistant, needle-like leaves (Fig. 5.08) that minimize exposure to the sun. Clusters of inconspicuous, small, white flowers emerge in the summer, when some of the older leaves are shed with increasing drought stress. Chamise has a high oil content, making it highly flammable and accounting for its nickname of greasewood.

There are few herbaceous plants in the understory of chamise chaparral, in part resulting from shade and competition from roots. Another hypothesized cause is allelopathy, the production of plant toxins that inhibit the germination and growth of competing plants. The importance of this phenomenon has been questioned in the light of evidence that seedlings may fail to mature because they are consumed by small herbivorous animals. When chamise is protected from herbivores by fenced enclosures, seedlings do develop and survive in appreciable numbers (Quinn, 1986).

FIG. 5.07. A. *Manzanita (Arctostaphylos) and California lilac (Ceanothus) chaparral in early spring at the Pinnacles National Monument in the Coast Ranges of Central California. B. California lilac chaparral with white blooms in early spring near Santa Barbara. C (opposite). Bigberry manzanita (Arctostaphylos glauca) at Pinnacles National Monument. This tall-growing species of the heather family (Ericaceae) has smooth red-brown bark and gray-green leaves.*

C

In Southern California, a close relative of chamise is red shank (*Adenostoma sparsifolium*), which may occur in pure stands or grow together with chamise (Hayes, 1977). Red shank is considered by some to form the most attractive type of chaparral. It grows as a shrub or as a small tree with multiple branches from the base. The feathery, chartreuse foliage is open enough to reveal the rust-red, shaggy branches. The open growth habit of red shank when compared to chamise allows a greater coexistence of other shrubs and herbs.

Ceanothus Chaparral. California lilac (*Ceanothus*) is represented by a multitude of species that grow as low shrubs to small trees with little leaf litter underneath. Its shiny green leaves create a dense shade, shutting out herbaceous understory plants. In spring, abundant clusters of blue, violet, or white flowers can cover entire hillsides (Fig. 5.07B). *Ceanothus* species have evolved a mutually beneficial association or symbiosis with nitrogen-fixing microorganisms living in root nodules, and are thereby able to thrive on nitrogen-poor soils where many other plants cannot compete.

Compared with other common chaparral shrubs, *Ceanothus* species are short-lived, with dead and dying plants leaving gaps after several decades. After a fire, several species, such as *Ceanothus crassifolius*, *C. oliganthus*, and *C. thyrsiflorus*, sprout from seeds, creating stands of a single age and species (Hayes, 1977). Other species associated with *Ceanothus*, such as chamise, scrub oak (*Quercus dumosa*), toyon (*Heteromeles arbutifolia*), and sugar bush (*Rhus ovata*), usually recover from a fire by stump sprouting.

FIRE

Many of California's most attractive suburbs are surrounded by chaparral, especially in the southern part of the state. The risk of destructive fires is greatest when the hot, dry Santa Ana winds come from the desert to the east and northeast. This happens most frequently in autumn, reversing the usual flow of moist winds from the ocean. Fires occur somewhere in the chaparral every year and cause the most costly damage near urban areas, where effective fire prevention has allowed a greater than natural buildup of dry brush.

Interestingly, fires in Southern California are larger, less frequent, and more destructive than those in neighboring Baja California, where fire prevention has been minimal. Under natural conditions, fires are less severe there because flammable brush has been kept in check by smaller fires. This reasoning has led to the use of prescribed burns on a limited scale to decrease the risk of later, uncontrolled fire damage.

Fires have always been a part of the natural cycle of regeneration of chaparral vegetation, but they have become more frequent and intense, particularly during the last century of urbanization. In San Diego County, which has the greatest area of chaparral of any county in California, summer thunderstorms are common. There are about 30 lightning-caused fires per year (Dunn, 1986), but today these account for only five percent of fires, and they rarely cause extensive damage. More than 90 percent of lightning-caused fires burn less than one-quarter acre. The remaining fires are caused by man and may be much more damaging. The worst fires are those that occur in 35- to 50-year cycles, when drought years come right after a series of wet years. These conditions result in a large buildup of dry, dead fuel such as in the severe fire years of 1928 and 1970.

The bright side of the chaparral's vulnerability to fire is the initiation of a cycle of renewal in its vegetation (Keeley and Keeley, 1986). Though the landscape looks like a charred wasteland after a fire, these conditions initiate a series of events that begin with the first substantial autumn rains (Table 5.01). Seedlings of annuals become established, and, in combination with bulbs, produce a spectacular display of wildflowers the following spring. Many of the chaparral shrubs that were dominant before the fire begin to sprout from root crowns or establish a new generation of seedlings. But they will take several years to approach their former size.

Among the plants that recover by sprouting (sprouters) are chamise, scrub oak, toyon, and coffeeberry (*Rhamnus* species). Plants that produce a new generation from seeds (seeders) include many species of *Arctostaphylos* and *Ceanothus*. With the notable exception of chamise, which does both, plants follow either one strategy or the other. In the two to four years after a fire, while the previously dominant shrubs are still small, herbaceous plants become dominant. Some of the more familiar ones include golden yarrow (*Eriophyllum confertiflorum*), matilija poppy (*Romneya coulteri*), larkspur (*Delphinium* spp.), peony (*Paeonia californica*), mariposa lily (*Calochortus* spp.), soap plant (*Chlorogalum* spp.), and blue-eyed grass (*Sisyrinchium bellum*). However, after five to ten years, taller chaparral shrubs reestablish a dense cover that crowds out most of the herbaceous plants.

FIG. 5.08. *Chamise* (Adenostoma fasciculatum) *has the widest distribution of any shrub of the chaparral. Its water-conserving leaves are needle-like. This member of the rose family has clusters of tiny, inconspicuous white flowers with five petals. (Drawing by Ràfols, 1977, reproduced with permission of the Institute of Ecology.)*

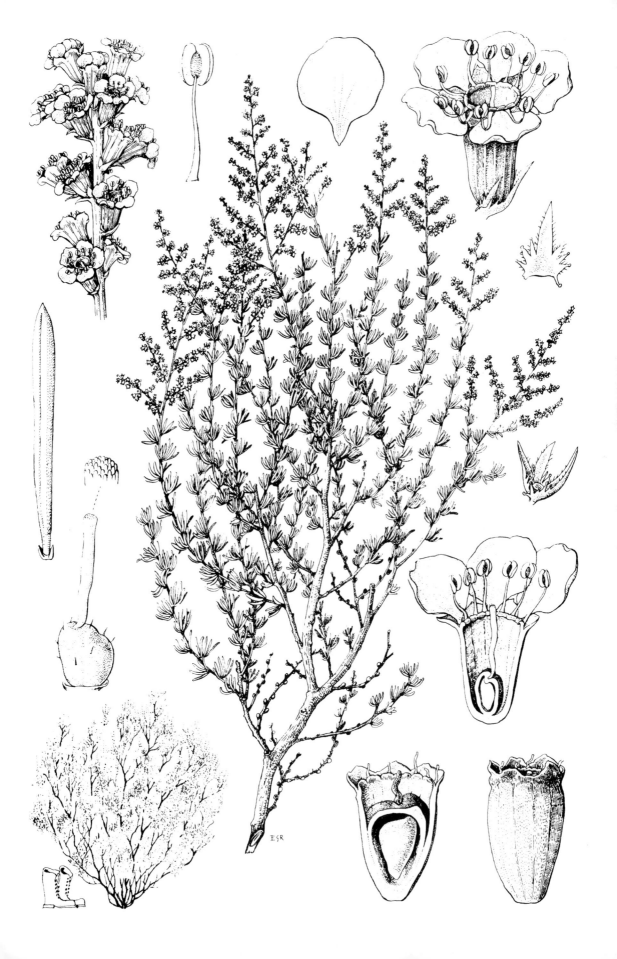

TABLE 5.01. THE CYCLE OF VEGETATION FOLLOWING A CHAPARRAL FIRE IN SOUTHERN CALIFORNIA

Years After Fire	Response in Vegetation
1	**Explosive growth of long dormant seeds and bulbs:**
	Annuals such as fire poppy (*Papaver californicum*); some with heat-stimulated seeds, such as whispering bells (*Emmenanthe penduliflora*).
	Bulbs such as mariposa lily (*Calochortus* spp.), *Brodiaea* spp., and star lily, death camas (*Zigadenus* spp.).
	Perennial herbs such as larkspur (*Delphinium* spp.) and figwort (*Scrophularia californica*) appear.
	Sprouter and seeder shrubs slowly begin to be re-established:
	Shrubs such as chamise (*Adenostoma fasciculatum*) and scrub oak (*Quercus dumosa*) begin to sprout from root crowns.
	Heat-stimulated seeds from non-sprouting *Ceanothus* and manzanita (*Arctostaphylos*) germinate.
2 - 4	**Herbaceous perennial shrubs become more conspicuous.**
	Examples: black sage (*Salvia mellifera*), golden yarrow (*Eriophyllum confertiflorum*), matilija poppy (*Romneya* spp.).
5 - 9	**Shrubs gradually become dominant, leaving less space for other plants.**
10 - 50	**Dense shrub cover with scant understory becomes increasingly vulnerable to wildfires in late summer and early fall.** Particularly destructive fires occur every 35 -50 years, when drought follows several wet years, resulting in buildup of abundant dry, dead fuel.
>25 - 50	**Decadence or senility occurs when thin layers of foliage are on the surface of shrubs, which consist largely of a thicket of dead branches.**

(From Keeley and Keeley, 1986.)

Manzanita Chaparral. Manzanita (*Arctostaphylos*), like California lilac, ranges from low shrubs to small trees (Figs. 5.07C, 5.09). Manzanita belongs to the heather family (Ericaceae), and of about 50 species of *Arctostaphylos*, more than 40 are found in California. Several of these species are endemic to small areas, and some are endangered. The Spanish name, manzanita or little apple, comes from the small red fruits that appear in summer and fall. In some species, leaves are oriented vertically to the sun to avoid the

FIG. 5.09. *Manzanita, meaning little apple in Spanish, is named for its small red fruit. The illustration shows bigberry manzanita (*Arctostaphylos glauca*), which has small, bell-shaped, pink to white flowers. (Drawing by Ràfols, 1977, reproduced with permission of the Institute of Ecology.)*

drying effect of direct sunlight. Small, bell-shaped pink or white flowers appear as early as December or January. Pure stands of manzanita form the densest of all types of chaparral. Only about half of manzanita species are able to resprout from stumps. Others reestablish themselves from abundant seeds that are stimulated to germinate after a fire.

Scrub Oak Chaparral. Scrub oak chaparral is dominated by evergreen oaks that grow as tall shrubs or small trees (Sawyer and Keeler-Wolf, 1995). Scrub oak (*Quercus dumosa*) (Fig. 5.11) typically has a thick canopy that reaches the ground. Underneath is abundant leaf litter, but there are few if any understory plants. However, scrub oak chaparral has a much greater diversity of associated species than most other forms of chaparral (Hayes, 1977). These species include the chaparral whitethorn (*Ceanothus leucodermis*), California holly or toyon (*Heteromeles arbutifolia*), and sugarbush (*Rhus ovata*). Scrub oak chaparral is found in relatively moist locations. In Southern California, it is typically on north-facing slopes. A quantitative survey of plant distribution showed that scrub oak was second only to chamise in the amount of cover in the chaparral of Southern California (Steward and Webber, 1981). After a fire, scrub oak chaparral recovers by resprouting rather than from seedlings.

Coastal Scrub Communities

Coastal scrub communities thrive in weather that is strongly influenced by the ocean. These areas extend inland for only a few miles where coastal ridges form a barrier, but penetrate more than 25 miles (40 km) where there is a break in the coast range, as at the Golden Gate at San Francisco and the San Fernando and San Bernardino valleys that extend inland from Los Angeles. Undeveloped parts of these regions are covered largely by low, dense scrub. Although coastal scrub appears similar to chaparral from a distance, it typically grows no higher than five feet (1.5 meters). Coastal scrub is referred to as soft chaparral because of the soft texture of its foliage.

In coastal scrub, something is in bloom almost the entire year, but the greatest abundance of flowers comes in late winter and spring. Many genera and species of coastal scrub plants with showy flowers are also found in chaparral, woodland, grassland, and clearings in forest. In late winter and early spring, fuchsia-flowered gooseberry (*Ribes speciosum*) and red-flowering currant (*R. sanguineum*) bloom (Fig. 5.12A and B). In mid-spring, patches of grassy meadows become brightly colored with many wildflowers, including Douglas iris (*Iris douglasiana*) and California poppy (*Eschscholzia californica*) and scarlet monkey flower (*Mimulus cardinalis*) (Fig. 5.12C-F).

FIG. 5.10. A. *Common flannel bush* (Fremontodendron californicum) *has large, bright yellow flowers in late spring and early summer. B. Farewell to spring (Clarkia rubicunda). C. Redbud (Cercis occidentalis) blooming in early spring. D. Late spring buckwheat (Eriogonum latifolium) with California lilac (Ceanothus spp.) and manzanita (Arctostaphylos spp.) in the background. E. California lilac (Ceanothus 'Gentian Plume'). A., B., and D., were photographed at Strybing Arboretum in San Francisco, C. and E. at the Santa Barbara Botanic Garden.*

A

B

C

D

E

Northern Coastal Scrub. From the Oregon border to the Big Sur coast, moist areas of coastal scrub are often dominated by almost pure stands of coyote brush (*Baccharis pilularis*) or bush lupine (*Lupinus arboreus*) (Ornduff, 1974). Among clumps of shrubs are grassy meadows with abundant wildflowers in spring.

Coastal Sage Scrub. Coastal sage scrub extends from just north of San Francisco to the northern part of Baja California. It is dominated by California sagebrush (*Artemisia californica*), black sage (*Salvia mellifera*) and purple-leaved sage (*S. apiana*), and/or white-leaved sage (*S. leucophylla*) (DeSimone, 1995). Coastal sage scrub is not as dense as chaparral and has more open space between shrubs, allowing room for herbaceous plants. Most of its shrubs regrow from seeds, and as a result there are typically shrubs of varying ages.

The soft leaves of coastal sage scrub are not as drought tolerant as the tough, thick leaves of the chaparral. Instead, many shrubs are drought-deciduous or seasonally dimorphic (page 33) (Westman, 1982, 1983). The drought-evasive adaptation of summer leaf loss is also prominent in the coastal vegetation of Chile and the Mediterranean Basin.

Fire frequency is as great in coastal sage scrub as it is in chaparral: about once every 20 years (Westman, 1982). This may be because the volatile oils that give the coastal sage scrub its fragrance are also flammable. After a fire, the most widespread shrubs, such as California sagebrush, can both resprout and produce abundant seeds which result in a large crop of seedlings the following year. The herbaceous plants, most of which are annuals, emerge from a dormant seed pool, stimulated by heat, charred wood, and light (O'Leary, 1988). Many of these fire followers are the same species that are found after a fire in the chaparral. In the coastal sage scrub, however, they continue to coexist with the larger shrubs and may help to recolonize adjacent areas of chaparral after a fire.

Oak Woodlands

The hilly oak woodlands of California comprise one of its most familiar and appreciated landscapes (Peattie, 1991; Bakker, 1984; Faber, 1990; Pavlik et al., 1991). California has 18 species of oak, of which nine are trees with tall, thick trunks. The remainder are scrub oaks, such as California scrub oak (*Quercus dumosa*), which dominates one type of chaparral (Fig. 5.11). Tree oaks dominate a landscape more conspicuously. They dot hillsides as scattered trees against a background of grasses that are green in winter and straw-colored in summer, and they form denser groves in the clefts of hills and in protected valleys. Oak woodlands and forests cover about 7.4 million acres (3 million hectares), or four percent of California's total area, almost as much as is occupied by chaparral.

Although oak trees are native to California, most of the grasses that grow with them were unintentionally introduced from the Mediterranean Basin. After arriving, primarily as contaminants of grain seeds, they spread rapidly, crowding out native bunch grasses.

FIG. 5.11. *Scrub oak* (Quercus dumosa). *(Drawing by Ràfols, 1977, reproduced with permission of the Institute of Ecology.)*

Fɪɢ. *5.12. A., B. Two red-flowering California species of* Ribes *that bloom in late winter and early spring. Fuchsia-flowered gooseberry (R.* speciosum*) (A.) is found near the coast from Central California south to Baja California. Pink winter currant (R.* sanguineum*) (B.) is native to the Coast Ranges but grows as far north as Canada. C. Douglas iris (*Iris douglasiana*). D. California poppy (*Eschscholzia

Grain was grown in California after the founding of the California missions in 1769 and on an increasing scale after the wave of settlement following the gold rush of 1849. Oak woodlands became ideal grazing lands for cattle and sheep and largely remain used for that purpose today. About 80 percent of oak woodland is currently in private hands, and over 75 percent is used for grazing (Ewing, 1990) (Fig. 5.13A).

Many California oaks are evergreen, such as coast live oak (*Quercus agrifolia*) (Fig. 5.13 B and C) and canyon oak (*Q. chrysolepis*). Others lose their leaves in the winter, creating dramatic silhouettes of dark trunks and branches against green hillsides. These include valley oak (*Q. lobata*), black oak (*Q. kelloggii*), and blue oak (*Q. douglasii*) (Fig. 5.13A and D). Engelmann oak (*Q. engelmannii*) can be considered intermediate because it may lose most of its leaves in a severe summer drought. The leaves of the deciduous oaks are thinner and less stiff than those of the evergreen species.

The flowers of all California oaks bloom with or after the appearance of new leaves in spring. Oaks are wind pollinated. They must therefore produce an enormous amount of pollen because most of it never reaches its destination. The pollen comes from long, pendulous clusters of up to 100 male flowers, called catkins. The female flowers occur singly and are less conspicuous. After pollination, some acorns mature within the same year (blue oak), whereas others develop over a two-year period (black oak).

To identify an oak tree, it is helpful to know what species are common in the area and to pick up acorns to examine their distinctive shapes (Fig. 5.15). The surface and branching pattern of the trunk are also helpful. Leaf shape can be confusing because in some species it will vary substantially in the same tree. The coast live oak, for example, has leaves that are specialized for either sun or shade. The sun-exposed leaves are small, thick,

D   E

californica), *the state flower. E. Spring flowers in the California native gardens at Strybing Arboretum in San Francisco. Included with Douglas iris and California poppy are white and pale yellow meadow-foam* (Limnanthes douglasii) *and pink checkerbloom* (Sidalcea malvaeflora). *F. Scarlet monkey flower* (Mimulus cardinalis).

and convex, and are able to trap sunlight in their several layers of photosynthetic cells. Shade leaves are flat, thin, and broad, and are better able to trap light where it is less intense. Other trees may defy straightforward identification because oaks commonly hybridize, developing characteristics intermediate between those of their parents.

Coast Live Oak. Coast live oak (*Quercus agrifolia*) is an evergreen tree that is characteristic of the landscape of the coastal plains, valleys, and foothills lying within about 50 miles (80 km) of the ocean. In Southern California, coast live oak grows to an elevation of about 5,000 feet (1,500 m). It is the only oak that thrives close to the coast, where it is often draped with hanging lichen. Coast live oak is typically found associated with bay laurel (*Umbellularia californica*), madrone (*Arbutus menziesii*), toyon (*Heteromeles arbutifolia*), California buckeye (*Aesculus californica*), or other oaks.

The Spanish called coast live oak *encina*, the same name used for the holm oak (*Q. ilex*) of the Mediterranean Basin. Holm and ilex both mean holly, which would also describe coast live oak with its spiny-edged, evergreen leaves. The tree characteristically branches into large, multiple trunks, some growing almost horizontally (Fig. 5.13B). The sinuous, twisting branches of some old trees hang close to the ground. The trees have a strange, enchanted appearance that was noted by early visitors to California. Coast live oaks often nestle into the contours of canyons and coastal hills, surrounded by chaparral or grassland (Fig. 5.13C).

Canyon Oak. Canyon oak (*Quercus chrysolepis*) is notable for having the widest distribution of any California oak. It is found throughout the entire length of the state, from sea level to 9,000 feet. Visitors to Yosemite Valley can see groves of canyon oaks below Yosemite Falls and the granite mass of El Capitan. In open areas, the trunks divide near the base, and the canopy forms a dome-shaped hemisphere that may trail to the ground at the periphery. Early leaves are small, thick, and holly-like, with a toothed margin, similar to those of the coast live oak. Later leaves are smooth. The undersurfaces of the young leaves and the acorn cups have a golden, felt-like covering, which was the basis of one of its common names, gold scale. The acorns are small and egg-shaped. They emerge from a cup that is broader than the acorn. The wood of the canyon oak was known to the pioneers for its strength and hardness. It was made into wedges with iron-rimmed heads that were used to split redwood for construction purposes. The wood was also prized for axles, wheels, and poles (Peattie, 1991).

Engelmann Oak. Engelmann oak (*Quercus engelmannii*) has the smallest range of any of the California oaks. It grows in Southern California, primarily in mid elevations of San Diego and Riverside counties, near two of the fastest-growing urban areas in the country (Scott, 1990). It is a partly drought-deciduous tree, losing many of its leaves during the summer dry season, particularly after winter rains have been scant. It requires fairly rich soil. The tree is medium sized, with small, smooth-edged, oval leaves. The acorns are oval and about half enclosed by a warty-surfaced cup (Fig. 5.14).

Engelmann oak is one of three species of California oak that are not reproducing well. The other two are valley oak and blue oak. In all three cases, there are many areas of large, old trees, often with many nearby seedlings. However, more established saplings and small trees are scarce, raising concerns about the long-term prospects for these species. Possible causes of a scarcity of saplings include heavy grazing by livestock and deer and spring competition for moisture from introduced annual grasses, which dominate or replace the native bunch grasses in most areas. Whatever the causes, there are now efforts to protect several areas from excessive grazing and to replant with locally grown acorns, seedlings, and saplings (McCreary, 1990).

Valley Oak. Valley oak (*Quercus lobata*) is the largest oak in North America, reaching heights in excess of 100 feet (30 m) and trunk diameters of six to nine feet (1 to 3 m). The rounded crown often has peripheral branches that droop nearly to the ground. It is a deciduous tree with leaves that are deeply lobed and felt-covered. Until the late 19th century, these grand trees dominated the landscape of the Central Valley, where they thrived in the deep, fertile, and seasonally well-watered soil. Now the most impressive remaining speci-

FIG. 5.13. A. *Cows grazing under blue oak* (Quercus douglasii) *in early spring. Seventy five percent of California's oak woodland is used for grazing. B. Coast live oak* (Q. agrifolia) *in the Santa Barbara Botanic Garden. Old trees such as this one have almost horizontal, sinuous branches that approach the ground. C. Coast live oak at the base of a canyon surrounded by chaparral on the slopes at Pinnacles National Monument in Central California. D. North slope-south slope variations in vegetation: an air view of an inner Coast Range landscape, showing dense growth of blue oaks only on the northern slopes. (Photograph by David Cavagnaro, reproduced with his permission and that of* Fremontia *[18(3):84, 1990].) E. Shooting stars* (Dodecatheon sp.) *and F. lupine* (Lupinus sp.) *growing in oak woodland.*

A

B

C

D

E

F

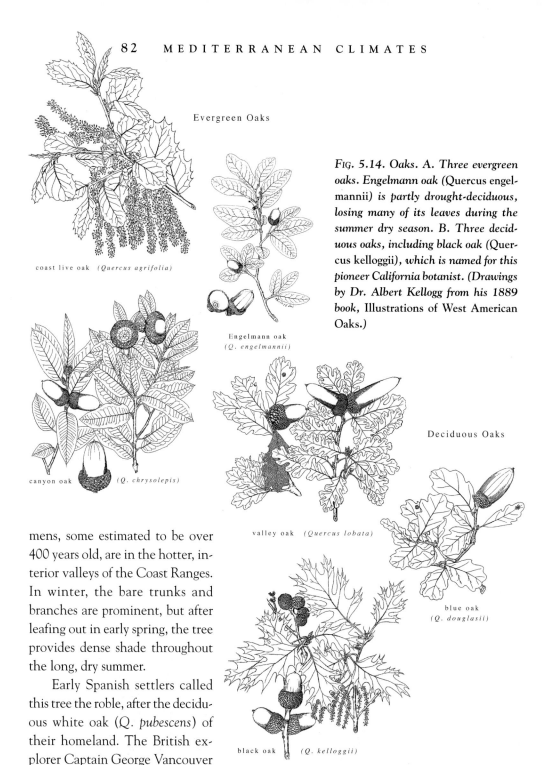

Evergreen Oaks

coast live oak   *(Quercus agrifolia)*

canyon oak   *(Q. chrysolepis)*

Engelmann oak
(Q. engelmannii)

valley oak   *(Quercus lobata)*

black oak   *(Q. kelloggii)*

Deciduous Oaks

blue oak
(Q. douglasii)

FIG. **5.14.** *Oaks. A. Three evergreen oaks. Engelmann oak (*Quercus engelmannii*) is partly drought-deciduous, losing many of its leaves during the summer dry season. B. Three deciduous oaks, including black oak (*Quercus kelloggii*), which is named for this pioneer California botanist. (Drawings by Dr. Albert Kellogg from his 1889 book,* Illustrations of West American Oaks.*)*

mens, some estimated to be over 400 years old, are in the hotter, interior valleys of the Coast Ranges. In winter, the bare trunks and branches are prominent, but after leafing out in early spring, the tree provides dense shade throughout the long, dry summer.

Early Spanish settlers called this tree the roble, after the deciduous white oak (*Q. pubescens*) of their homeland. The British explorer Captain George Vancouver also revealed his homesickness when he visited the Santa Clara Valley in 1796 and had the impression that its park-like landscape had been "planted with the true English oak." The acorns of the valley oak were used as a staple food by the Indians. They have a distinctive cartridge-like, elongated shape that comes to a sharp point, set into a warty-surfaced cup. Fremont, in his famous1844 expedition to California, obtained a half bushel of its roasted acorns by trade and found them sweet and tasty.

The most famous valley oak, until it succumbed to a 1977 storm, was the giant "Hooker oak" in Chico, California. It was named in honor of Sir Joseph Hooker, the famous botanist and director of the Royal Botanic Gardens at Kew, near London, England. Hooker visited the tree in 1877, in the company of the California naturalist John Muir and the American botanist Asa Gray. He considered it to be the largest oak in the world, though it turned out not to be. Valley oaks have disappeared rapidly since that time, even though their wood was considered useless except for fuel, "hard but brittle; …heavy but weak" (Peattie, 1991). Trees were lost primarily because of their preference for rich, well-watered bottomland, which resulted in their being cleared for agriculture and urbanization. Other trees died from a lack of ground water due to increased well irrigation and stream diversion.

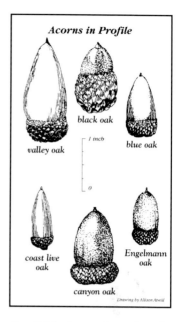

FIG. 5.15. Acorns from the oaks illustrated in Fig. 5.14 shown for comparison. The distinctive shapes of acorns are helpful in identifying the species of an oak. (Reproduced with permission from Pavlik et al., 1991.)

Black Oak. Black oak (*Quercus kelloggii*) is named after Dr. Albert Kellogg, a pioneer botanist whose drawings of California oaks were published just over a century ago; some are reproduced in this chapter (Fig. 5.14). Black oak is found in mountain areas to an elevation of 7,000 feet (2,100 m), where rainfall is relatively abundant and where frost and snow may be common. The most extensive stands are in the northern half of the state, both in the foothills of the Sierra Nevada and in the Coast Ranges, but away from the coast itself.

Black oak supplied what were generally considered to be the best-tasting acorns for the Indian's diet. The acorns are also prized by the redheaded California woodpecker, which hammers nuts so deeply and tightly into holes it has made in the trunk that they are impossible for squirrels to dislodge. Acorns are more than halfway covered by their cups, which together make an oval shape. The deciduous leaves are deeply divided with angular lobes that have long, pointed tips. In the spring, new foliage is bright pink or crimson for a few weeks, and in the fall, the leaves become a vivid yellow.

Unlike many other oaks, black oak does not dominate its landscape. It grows near other broadleaf trees and conifers, often together with canyon oak (*Q. chrysolepis*), madrone (*Arbutus menziesii*), ponderosa pine (*Pinus ponderosa*), white fir (*Abies concolor*), incense-cedar (*Calocedrus decurrens*), and sugar pine (*Pinus lambertiana*).

Blue Oak. Blue oak (*Quercus douglasii*) is a deciduous oak that is notable for being able to thrive under extreme conditions. It tolerates shallow soil, high summer temperatures, and relatively arid conditions. A thick coating of wax on the leaves minimizes water evaporation and gives the trees their bluish color. In unusually dry years, the trees conserve water further by losing many of their leaves in late summer and early fall.

Blue oaks grow below 3,500 feet (1,050 m) in the Coast Ranges and in hot, dry interior foothills of the Sierra Nevada. They are abundant along California Highway 49. Although they typically grow as well-separated trees in park-like stands, groups may cluster together to form a single large canopy on north-facing slopes (Fig. 5.13D). Their branches are unusually brittle, and winter storms create a litter of broken limbs under trees.

Forest

Coast Redwood. Of California's many types of forest, coast redwood (*Sequoia sempervirens*) forests are the most visited and best known. Not only do they include the world's tallest trees, several that are in excess of 350 feet (105 meters), they also attain an impressive *average* height of well over 200 feet (60 meters) in protected canyon areas (Fig.

5.16A). Under favorable conditions, they can reach that size within the relatively short time of less than a century, being among the world's fastest growing trees. The coast redwood is also a very long lived tree by virtue of its resistance to fire and disease. An age in excess of 500 years is common, and some individual trees are over 2,000 years old.

Coast redwoods have an ancient lineage, dating back more than 160 million years, to early in the age of the dinosaurs. Fossils show that redwoods once covered much of

A   B

FIG. 5.16. A. *Coast redwood* (Sequoia sempervirens) *at Humboldt State Park, north of San Francisco. B. The two kinds of foliage that grow on mature coast redwood trees are shown photographed against the tree trunk. The bottom two-thirds of the tree has narrow, flat, pointed needles that grow out from the branchlet in one plane (left). On the top third of the tree, more exposed to the sun, is the so-called sun foliage, consisting of short, overlapping, scale-like leaves that cover all sides of the branchlet (right). These branchlets are almost identical to those of the coast redwood's close relative, the giant sequoia* (Sequoiadendron giganteum), *which grows on the western slope of the Sierra Nevada.*

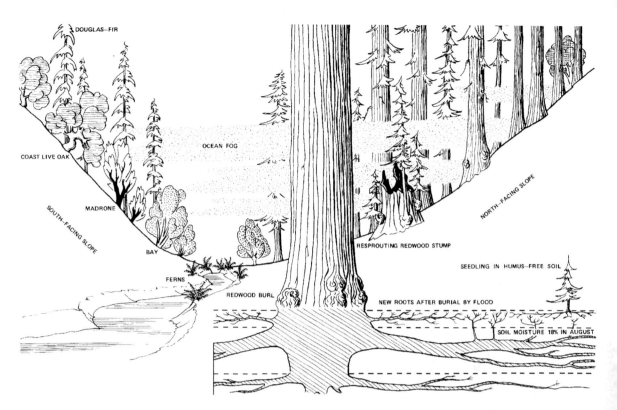

FIG. 5.17. *A foggy canyon with coast redwood near the stream and on its adjacent, north-facing slope. The sunnier, south-facing slope has Douglas fir, coast live oak, madrone, and California bay trees, all of which are more drought tolerant than coast redwood. The drawing illustrates the shallowness of the main horizontal root system of the coast redwood. New horizontal roots grow when the main root system has become more deeply buried by soil deposited by the overflowing stream during winter storms. (Drawing by Gerhard Bakker from Elna Bakker's* An Island Called California, *reproduced with permission, University of California Press.)*

North America, Europe, and Asia. The present, much-restricted range of coast redwood straddles the moist end of the mediterranean climate region of California and extends into the southernmost portion of the temperate rainforest of the Pacific Northwest. The annual rainfall where coast redwoods occur is about 35 to 80 inches (90 to 200 cm). Winter rain is augmented by as much as 10 inches (25 cm) of fog drip falling mainly between May and September. This fog drip makes it possible for redwood forests to retain critical soil moisture levels during August, the driest month. Roots grow close to the surface and spread widely and thus are well adapted to take advantage of this light summer precipitation.

Coast redwoods are confined to a narrow coastal fog belt, extending north from the Big Sur coast, past San Francisco, to their largest stands in the three northernmost coastal counties, Del Norte, Humboldt, and Mendocino (Fig. 5.04), and into the extreme southwestern corner of Oregon. They avoid the salt spray next to the coast and start at least a mile (1.6 km) inland, extending no further than 45 miles (72 km) from the coast in the northern part of the state. The width of this band decreases toward the southern part of the tree's distribution.

Even within this limited range, coast redwoods are further isolated in foggy, wind-sheltered canyons, river bottoms, and north-facing slopes (Fig. 5.17). In these cooler, moister

FIG. 5.18. A. *Monterey pine* (Pinus radiata) *and Monterey cypress* (Cupressus macrocarpa) *on the cliffs above the Pacific Ocean at Point Lobos State Park near Carmel, south of San Francisco on the coast of Central California. B. Cones of the Monterey pine remain tightly closed (opposite) until the heat from a fire or an exceptionally hot day opens them.*

settings, coast redwoods dominate their habitat to a remarkable degree, creating their own microclimate in forests that have relatively few understory species. Redwood trees throw a dense shade and drop a thick layer of plant debris, which reduces competition from other plants. Coast redwood forests are easy to walk through because of the relative lack of underbrush, except where trees have fallen. The cool, hushed, and serene atmosphere and the moist, fragrant, and resilient duff underfoot make a walk in the redwoods a pleasant experience. Redwood forests are particularly inviting in the summer, when their shade and moist air provide a refreshing drop in temperature.

Redwood groves are often intermingled with patches of mixed evergreen forest, oak woodland, or chaparral. Chaparral and broad-leaf, sclerophyll evergreen trees favor hot and dry south-facing slopes. Mixed evergreen forests increase in extent and are gradually supplanted by oak woodlands in the drier inland areas that extend toward the Central Valley. Rather than representing an orderly sequence, all of these differing plant communities make up an attractive and complex mosaic of vegetation.

Mature coast redwood trees are notable for having two kinds of foliage. The bottom two-thirds of the tree, which is all that is low enough to be visible from the ground, has narrow, flat, pointed needles that grow in one plane, straight out from a branchlet. The number of nodes along a stem indicates the number of growing seasons. On the top third of trees, where branches are most exposed to the sun, foliage consists of short, overlapping, scale-like needles that cover all sides of the branchlet. These branchlets, often knocked to the ground by a severe winter storm, seem as though they come from another kind of tree. Their foliage is almost identical to that of the coast redwood's close relative, the Sierra redwood or giant sequoia (*Sequoiadendron giganteum*). The giant sequoia grows on the western slope of the Sierra Nevada and has the most massive trunk of any species of tree.

The root system of the coast redwood is surprisingly shallow for such a tall, heavy tree. But it compensates by being unusually broad, reaching a radius of as much as 100 feet (30 meters), particularly in the case of the tallest trees along streams. Roots are able to tolerate low levels of oxygen during periodic winter flooding. When the root system becomes more deeply buried under layers of soil deposited by an overflowing stream, a new horizontal layer of roots grows just under the higher ground surface. This gives the streamside tree a multi-storied root system, the top layer of which can take full advantage of fog drip during the summer dry season (Fig. 5.17).

The survival of larger tracts of second-growth redwood forests despite years of extensive logging can be attributed to the tree's unusual regenerative capacities. When the parent tree is cut or damaged, small sprouts from burls at the base of the trunk grow within several decades into tall, second-growth forests. These young trees often reveal their origins as stump sprouts by growing in a circle.

Even though redwood forests retain moisture during the summer, they are subject to forest fires, particularly at the end of the dry season. Indeed, redwoods are fire-adapted like many chaparral plants. Their seedlings require partial sun and will not sprout except in newly fire- or storm-exposed soil. Before European settlement, fires in a redwood grove are believed to have occurred four or five times per century. Thick, insulating bark and a lack of resin make coast redwoods much more fire-resistant than the resinous and more flammable Douglas-fir. Nevertheless, it is common to see older trees hollowed out or otherwise extensively blackened and scarred by fire.

B

Although coast redwoods often form relatively pure groves, other trees can be close associates. Among the largest of these is Douglas-fir (*Pseudotsuga menziesii*). Its range extends far beyond that of the coast redwood, north into British Columbia, east to the Rocky Mountains, and south into Mexico. Douglas-fir is now a major source of lumber for the world, most of it coming from Oregon, Washington, and British Columbia.

Smaller evergreen trees that grow within or on the edge of coast redwood groves are California bay (*Umbellularia californica*), madrone (*Arbutus menziesii*), and tanbark oak (*Lithocarpus densiflora*). California bay is noted for its long, shiny, pointed leaves that are extremely pungent when crushed. They are often used in stews as a substitute for true bay leaves (*Laurus nobilis*), which come from the Mediterranean Basin. Storms often bend these trees into a permanently near-horizontal position from which the trunks sprout vertical stems.

Madrone is notable for its smooth reddish-brown bark that peels in thin layers. It has a broad range, extending from southern California to British Columbia. Madrone is in the same genus as the strawberry tree (*Arbutus unedo*), a native of the Mediterranean Basin. Both have pale, bell-shaped flowers typical of the heath family. They are also noted for their colorful fruit that becomes prominent in the fall, yellow and orange on the madrone and yellow, orange, and red on the strawberry tree.

Closed-Cone Conifers. The vast majority of conifers have cones that open at maturity to disperse their seeds by wind and gravity. However, about one-third of California's conifers, 18 species of pines and cypresses, retain their seeds in tightly closed cones that open only at high temperatures, typically after a fire. The most fire-dependent of these species is probably the knobcone pine (*Pinus attenuata*), whose cones open when a temperature of about 200°F (94°C) is maintained for five minutes (Barbour et al., 1993). Most closed-cone pines are found along the coast and include bishop pine (*Pinus muricata*) (Fig. 3.09), beach pine (*P. contorta*), and Monterey pine (*P. radiata*).

Monterey pine reaches an average height of about 80 feet (25 m). Its trunk is red-brown and deeply furrowed. Needles usually grow in bunches of three. Point Lobos State Park, just south of Carmel, is one of the most beautiful places to see these trees and is the only area where they grow together naturally with Monterey cypress (*Cupressus macrocarpa*), another closed-cone species (Fig. 5.18A).

Monterey pine is the most widely planted timber tree in the world but is highly restricted in its own natural range. It is found naturally in only three places in California, which add up to about 11,000 acres (4,400 hectares), with a little over a third in the form of natural forest (Deghi et al., 1995). In this limited distribution, well within the coastal fog belt, it favors well-drained sandy or loamy soil. The largest of its three natural sites is around the resort towns of Monterey and Carmel, where the remaining forests are threatened not only by the spread of housing and other development, but by pitch canker, a fungus disease that is transmitted by three species of beetle. This potentially lethal tree disease was found to have spread to California in 1986 and to the natural groves of Monterey pine on the Monterey Peninsula in 1992.

PLANTS FOUND IN MEDITERRANEAN CLIMATE AREAS OF CALIFORNIA

Adenostoma fasciculatum	chamise, greasewood	Rosaceae	evergreen shrub
Arbutus menziesii	madrone	Ericaceae	evergreen tree
Arctostaphylos spp.	manzanita	Ericaceae	evergreen shrub
Artemisia californica	coastal sagebrush	Asteraceae	shrub
Baccharis pilularis	coyote brush	Asteraceae	evergreen shrub
Ceanothus spp.	California lilac	Rhamnaceae	evergreen shrub
Epilobium californica	California fuchsia	Onagraceae	shrub
Eschscholzia californica	California poppy	Papaveraceae	herb
Fremontodendron californicum	flannel bush	Bombacaceae	shrub
Heteromeles arbutifolia	toyon, California holly	Rosaceae	shrub
Iris douglasiana	Douglas iris	Iridaceae	bulbous
Lupinus spp.	lupine	Fabaceae	herb/shrub
Mimulus spp.	monkey flower	Scrophulariaceae	shrub
Pinus radiata	Monterey pine	Pinaceae	evergreen conifer tree
Pseudotsuga menziesii	Douglas-fir	Pinaceae	evergreen conifer tree
Quercus agrifolia	coast live oak	Fagaceae	evergreen tree
Quercus chrysolepis	canyon oak	Fagaceae	evergreen tree
Quercus douglasii	blue oak	Fagaceae	deciduous tree
Quercus dumosa	scrub oak	Fagaceae	evergreen shrub
Quercus kelloggii	black oak	Fagaceae	deciduous tree
Quercus lobata	valley oak	Fagaceae	deciduous tree
Rhus ovata	sugar bush	Anacardiaceae	evergreen shrub
Salvia spp.	sage	Lamiaceae	perennial shrub
Sequoia sempervirens	coast redwood	Taxodiaceae	evergreen conifer tree
Umbellularia californica	California bay	Lauraceae	evergreen tree

With so little natural forest of Monterey pine left and much of this interspersed with urban development, the fire by which these trees normally reproduce becomes a matter of special concern. Natural forests of Monterey pine have a dense pattern of growth, with dead lower limbs that stay on the tree, making walking difficult. Dead branches together with flammable shrubs in the understory make fires in these forests very destructive and threatening to nearby housing. Fire management becomes a complex problem when fire is normally required to stimulate opening of the cones (Fig. 5.18B) and create favorable conditions for a new crop of seeds to germinate for replacement of trees whose lifespan is not much over 100 years.

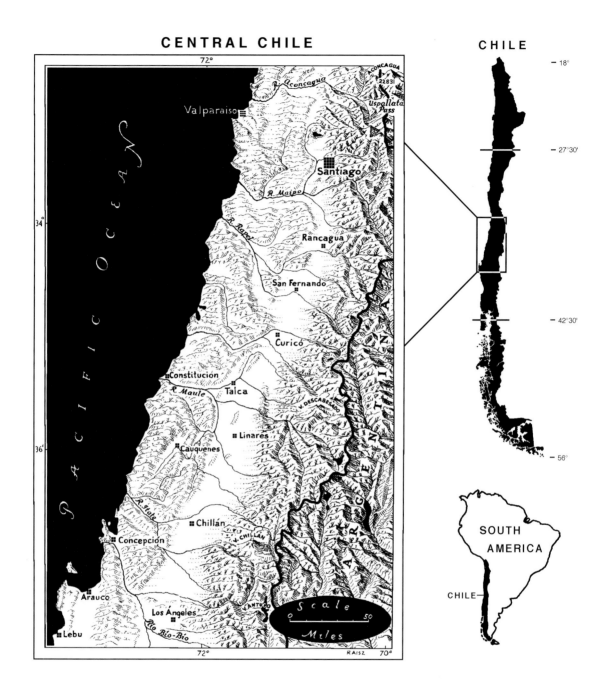

CENTRAL CHILE

CHILE

SOUTH
AMERICA

CHILE

FIG. 6.01. *The heart of the mediterranean climate area of Central Chile, with orientation maps showing the entire length of the country and the position of Chile in South America to the right; the two horizontal lines on the map of Chile show the portion of the country discussed in this chapter.*

The Coast Ranges form a longitudinal band along the Pacific Coast. Further inland is the Central Valley, which forms a rich agricultural belt and contains the capital city of Santiago. To the east are the high, snow-covered Andes Mountains. (Adapted from a map by Erwin Raisz with permission of John Wiley & Sons.)

6.
CENTRAL CHILE

LANDSCAPE AND CLIMATE

Chile is a uniquely long and narrow country that spans almost 40° of latitude (Fig. 6.01). Its northern boundary with Peru is more than 2,600 miles (4,200 km) from Cape Horn to the south. In terms of the northern hemisphere, this is equivalent to the distance and latitudes from Mexico City to Juneau, Alaska, or from the bend of the Nile in northern Sudan to Copenhagen, Denmark. In contrast to its extreme length, the east-west diameter of Chile averages only about 100 miles (160 km), a distance that can easily be seen from a plane. There is only one long major highway, and that runs in a north-south direction.

The narrow east-west dimension of Central Chile compared to California can be appreciated from the schematic cross sections of the two (Fig. 6.03). As in California, coast ranges rise to peaks as high as 7,250 feet (2,200 m). The Valle Central, or Central Valley, has a thriving agriculture and is the location of most of Chile's cities and towns. To the east lie the steep Andes, which rise to peaks of higher elevations than those of the Sierra Nevada.

By virtue of its geography, Chile is something of an ecologic island, surrounded by the Pacific Ocean to the west, the Atacama Desert to the north, and the stormy, icy waters of Drake Passage between it and the Antarctic continent to the south. The eastern border between Chile and Argentina is on the crest of the Andean mountain chain, which forms an almost continuous wall between the two countries. East of the capital, Santiago, the peaks of the Andes rise to about 21,500 feet (6,500 m), and the mountain passes are about 12,500 feet (3,800 m) or higher. Further south, the chain continues with peaks that gradually decrease in elevation and form a less formidable barrier, consisting of numerous active, snow-covered volcanoes with forested lower slopes. Chile's geographical isolation by desert, ocean, and mountains helps to account for the fact that so many of its plants are endemic (page 56) (Weber, 1990).

Central Chile, which includes both its mediterranean climate area and most of the population, extends for about one-third of the length of the country (Fig. 6.01). Its climate is strikingly similar to that of California (Thrower and Bradbury, 1973). The climate maps are shown next to those for California to facilitate comparisons (pages 60 and 61).

Rainfall

The driest areas are toward the equator and the wettest are toward the poles, as in California (Fig. 5.02). In most of both mediterranean climate regions, at least 85 percent of the precipitation falls in the winter half of the year. La Serena, at the northern extreme of the mediterranean climate region, is driest. Santiago, the only very large metropolitan area in Chile, is centrally located between the moister southern and the drier northern parts of the mediterranean climate region. Rainfall is in excess of 40 inches (100 cm) per year between Concepción and Puerto Montt, in the southern part of Central Chile.

As in California, the areas with the heaviest winter rains also have the shortest summer dry seasons, with fewer than four dry months in Concepción and more than eight in La Serena. Summer coastal fog is common along the Pacific coast as it is in California, providing some moisture and moderating the temperature during the summer dry season. When winter storms arrive, the pattern of precipitation is one of relatively infrequent but severe storms that often cause flooding. In the Andes, winter precipitation accumulates as snow, as it does in California's Sierra Nevada. As a result, rivers and streams have their highest average flow in the spring and early summer and are lowest in late summer and early autumn.

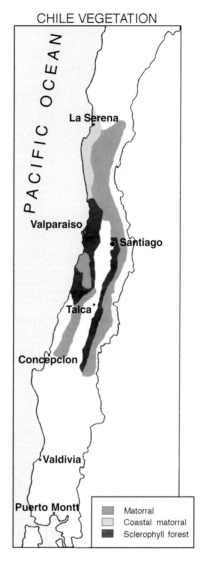

FIG. 6.02. Plant communities. matorral, coastal matorral, and sclerophyll forest. (Based on Gajardo, 1994.)

Temperature

Temperature along the coast is mild, with almost frost-free winters. There is relatively little difference between La Serena in the north and Concepción much further south (Fig. 5.03). The seasonal differences in temperature along the coast are also modest. Chile's Central Valley is less hot in the summer than California's Central Valley because the distance from the cold currents of the Pacific Ocean to the Andean slopes is much smaller. There are also many breaks in the coast ranges created by numerous rivers that run swiftly and along fairly straight courses from the Andes to the ocean. The Andes are high enough to block the passage of hot summer air and cold winter air from Argentina to the east. In addition, the temperature of the northern part of the Central Valley is also moderated by its slightly higher elevation compared with that of California: Santiago, for example, lies at an elevation of 1,700 feet (500 m) compared with Sacramento, which is close to sea level.

Plant Communities

As one travels from north to south in Chile, there are changes in vegetation similar to those seen along the California coast progressing north from Baja California and on to the Pacific Northwest, through British Columbia and the panhandle of Alaska (Fig. 6.04). From a latitude of 18° to about 28° S, Chile is a desert with vegetation becoming more abundant only where moisture levels are augmented by coastal fog or by ground water along the rare stream basin. Further south, from about 28° to 32°, is the semi-arid region referred to as *el Norte Chico*, the Little North. Near the coast, scrub vegetation reminds one of San Diego County and the northern part of Baja California, where the coastal fog also provides moisture and moderates temperatures. The most typical mediterranean climate belt occurs between about 32° and 37°. Here, matorral scrub and broadleaf evergreen woodland and forest correspond to California chaparral and oak woodland and forest between

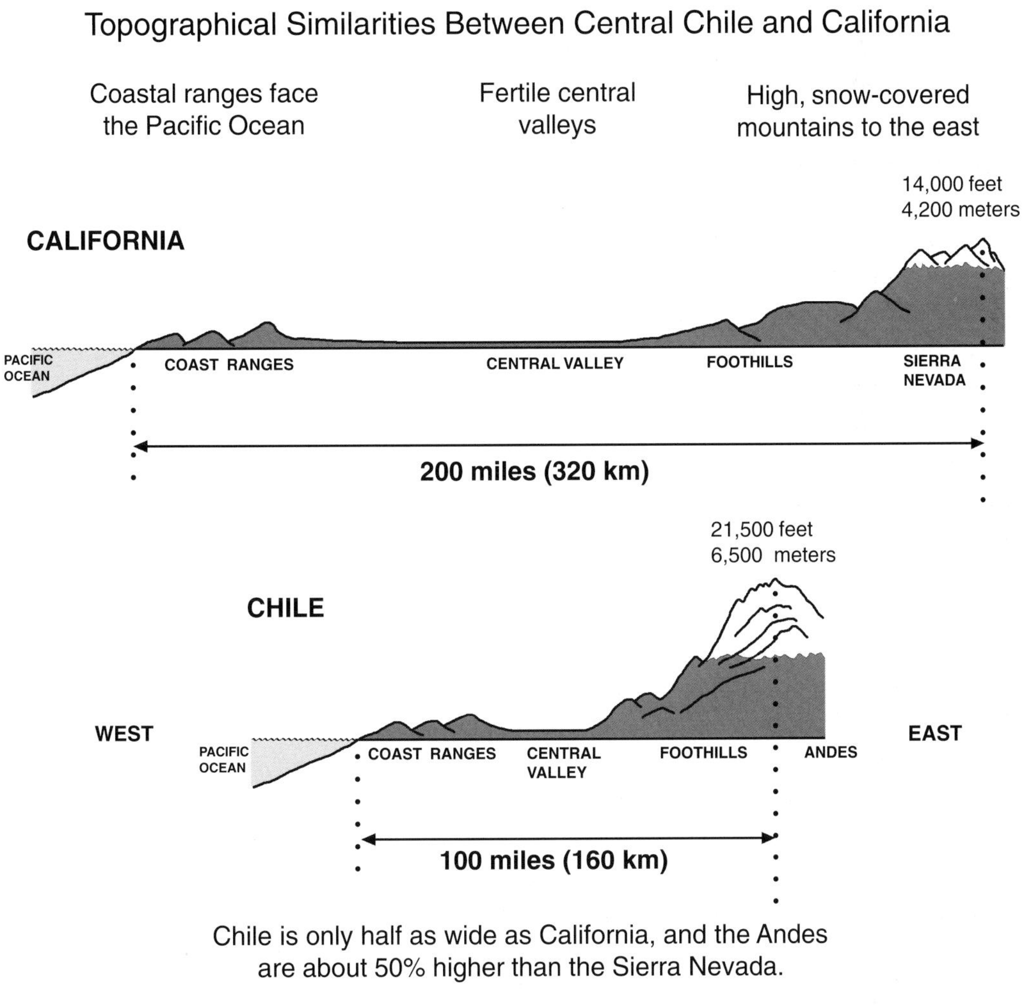

Topographical Similarities Between Central Chile and California

Coastal ranges face the Pacific Ocean	Fertile central valleys	High, snow-covered mountains to the east

14,000 feet
4,200 meters

CALIFORNIA

PACIFIC OCEAN COAST RANGES CENTRAL VALLEY FOOTHILLS SIERRA NEVADA

200 miles (320 km)

21,500 feet
6,500 meters

CHILE

WEST PACIFIC OCEAN COAST RANGES CENTRAL VALLEY FOOTHILLS ANDES **EAST**

100 miles (160 km)

Chile is only half as wide as California, and the Andes are about 50% higher than the Sierra Nevada.

FIG. 6.03. *Central Chile and California compared in cross section. The topographical similarities are striking, but there are also several differences.*

San Diego and San Francisco (Figs. 6.02, 6.04). Differences in vegetation between the coast ranges and Andean foothills are fewer than analogous differences in California, probably because the narrowness and milder climate of Chile's Central Valley is less of a barrier to plant migration (Arroyo et al., 1995).

In the southern part of Central Chile, large temperate forests begin south of the Río Bío-Bío at about 37°, inland from Concepción. The climate from there south to Puerto Montt near 42° of latitude corresponds roughly to the coastal region between San Fran-

FIG. 6.04. *Central Chile is shown sideways, with north on the left and south on the right. The vegetation of each region is shown against a scale in meters. As duration of the dry season decreases and the amount of rainfall increases from north to south, vegetation becomes taller, denser, and eventually more lush. The scant semi-desert vegetation around Copiapó in the north exists in a climate with more than ten dry months per year. The mediterranean climate area of Chile extends roughly from La Serena at its dry extreme to near Concepción, at the moist end. South of Concepción, the brevity of summer drought no longer strictly conforms to the criteria for a mediterranean climate. The natural vegetation becomes more lush, with tall trees containing epiphytes and entwined by vines. Many of the plants discussed in the chapter are listed above the region where they grow. The change from desert to rainforest occurs over a distance of about 750 miles or 1200 kilometers. (Adapted from a diagram by DiCastri, 1981.)*

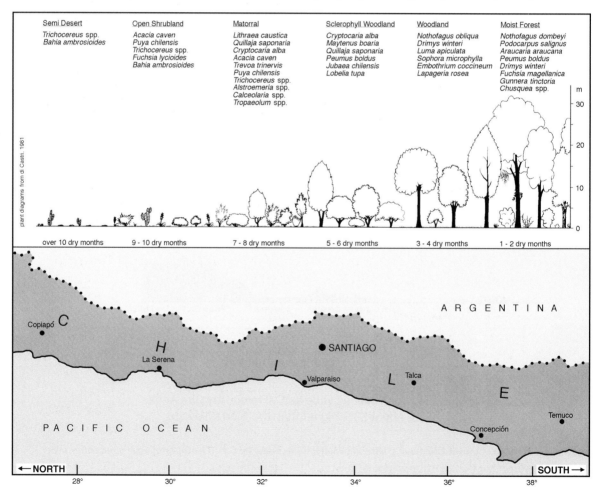

ELEVATION

The effect of elevation on vegetation is illustrated in Fig. 6.05. North to south changes near sea level are similar to those illustrated in Fig. 6.04, but the upper limit of vegetation in the Andes near La Serena is about 13,000 feet (4,000 m) and falls to about half that elevation in the southern part of Central Chile. Populations of mountain cypress (*Austrocedrus chilensis*) near Santiago, and monkey puzzle tree (*Araucaria araucana*) south of Concepción both occur at about 3,500 feet (1,000 m).

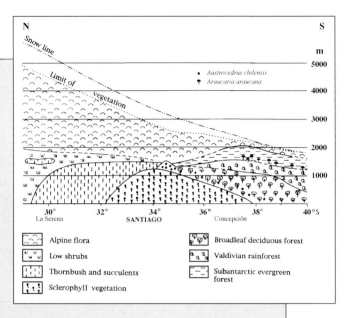

FIG. 6.05. *Vegetation at various elevations in Central Chile. The snow line and upper limits of vegetation descend from close to 16,400 feet (5,000 m) near La Serena in the north to just above 6,500 feet (2,000 m) inland from Concepción. Near Santiago, sclerophyll matorral and woodland extend to about 4,000 feet (1,250 m), where there are also woods of mountain cypress (Austrocedrus chilensis). Further south at the latitude of Concepción, sclerophyll vegetation is found mainly below about 3,000 feet (900 m) and broadleaf deciduous forest, mountain cypress, and Valdivian rainforest are at higher elevations. Groves of the distinctive monkey puzzle tree (Araucaria araucana) are near the tree line south of Concepción. (From Schmithüsen, 1956.)*

cisco and Seattle. It is wet but still has a mediterranean tendency, with one or two dry months a year in inland areas. Groves of the once abundant monkey puzzle trees (*Araucaria araucana*) are found in the coast ranges and the Andean foothills near Temuco. Near Puerto Montt is the home of the Fitzroy cypress or alerce (*Fitzroya cupressoides*). These species, together with mountain cypress (*Austrocedrus chilensis*) and several species of *Podocarpus*, are the major conifers of Chile. Although they had a much wider distribution before European settlement, conifer forests were never as extensive in Chile as in California. The southernmost third of Chile that lies beyond 42° latitude is cooler and rainier. It has a coastline forested by southern beech (*Nothofagus*) with numerous islands and fjords, like British Columbia and the panhandle of Alaska. In many ways, Chile is similar to the North American west coast but turned upside down.

As in other mediterranean climate regions, there are local variations in vegetation related to sun exposure, degree of protection from wind, and proximity to a stream. Fig. 6.06 illustrates variations in vegetation over distances of less than 165 feet (50 meters) in a small ravine or canyon (also called quebrada in Spanish). At the bottom, close to a small stream, are trees of medium height. A shady, more south-facing slope supports smaller trees

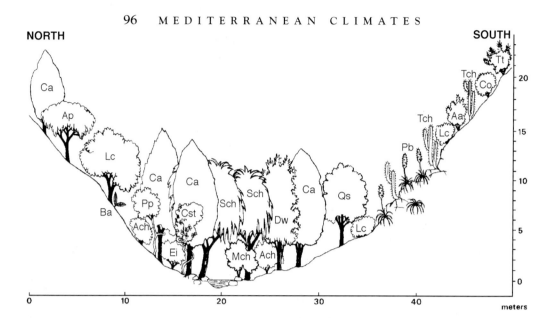

FIG. 6.06. *The vegetation in a ravine near the town of Tiltil, northwest of Santiago. The north-facing, south side of the ravine is more sun-exposed in the southern hemisphere. This is where the more drought-tolerant plants are found. Taller, more moisture-requiring shrubs and trees are on the south-facing, north side of the ravine and on the banks of the stream. The scale is in meters. (From Thrower and Bradbury, 1977.)*

Aa	Adesmia arborea	Ei	Escallonia illinita
Ach	Aristotelia chilensis	Lc	Lithraea caustica
Ap	Azara petiolaris	Mch	Myrceugenella chequen
Ba	Blechnum auriculatum	Qs	Quillaja saponaria
Ca	Cryptocarya alba	Pp	Proustia pyrifolia
Cst	Cissus striata	Sch	Salix chilensis
Co	Colliguaya odorifera	Tch	Trichocereus chilensis
Dw	Drimys winteri	Tt	Trevoa trinervis

and vines, while a sunny, more north-facing slope has *Puya* (bromeliad), cacti, and other more drought-tolerant plants.

Matorral

Matorral is the name given to shrubby sclerophyll vegetation in Chile. It extends from near the coastal city of La Serena to Concepción and the Río Bío-Bío, Chile's longest river. Matorral is concentrated in the coast ranges (Fig. 6.07B) and in the Andean foothills (Fig. 6.07E). Relatively little of it remains today in an undisturbed state.

Coastal matorral is a species-rich form of vegetation that resembles California's coastal scrub

A

C

B

D

communities in many respects and will be discussed separately. The transition from coastal matorral to matorral further inland is more gradual and imperceptible than from coastal sage scrub to chaparral in California (Fuentes et al., 1995).

Matorral shares many characteristics with chaparral of California, but it differs in some important respects (Thrower and Bradbury, 1977; Fuentes et al., 1986). Matorral shrubs tend to grow less densely than shrubs in the chaparral of California or maquis of the Mediterranean Ba-

E

sin. On the sunny, north-facing slopes and on more level areas with relatively scant rainfall, there are typically open spaces between clumps of shrubs (Fuentes and Muñoz, 1995), cacti, and terrestrial bromeliads. Only on the moister, shady, south-facing slopes is there a closed canopy of shrubs and low trees.

FIG. 6.07. *Matorral and espino. A (opposite). Acacia caven in bloom. B. Coast Range matorral in La Campana National Park. C. Litre (Lithraea caustica), a common matorral shrub or small tree, sprouts from its base after pruning or after a fire. D. Espinal landscape in the Central Valley just north of Santiago. Espinal is a sparser savanna landscape that is named after its dominant shrub, the thorny espino (Acacia caven). E. Matorral and streamside woodland in the Andean foothills in Río Clarillo National Reserve near the southern outskirts of Santiago.*

Another distinctive characteristic of matorral is its abundance of vines, bulbs, and other herbaceous plants, which grow primarily under the shelter of taller shrubs. A paucity of these herbaceous plants in open spaces between shrubs is attributed to voracious grazing by rabbits that were introduced from Spain about 50 years ago and that have become very abundant (Fuentes et al., 1983; Quinn, 1986). In contrast to the invasive rabbits, native herbivorous rodents graze lightly and stay under and near the protective cover of the shrubs to hide from predators. However, large predators such as condors have decreased in number and are not a great threat to the rabbit population. As a result, rabbits graze heavily in open areas. Taller shrubs shelter some of their own seedlings as well as herbs, but seedlings out in the open are more likely to be eaten. This vulnerability of seedlings may help to explain why clearings made in matorral are recolonized very slowly by surrounding shrubs and herbaceous plants (Fuentes et al., 1986). The open characteristic of matorral represents an example of a rapid change in a plant community resulting from introduced animals.

Litre. Litre (*Lithraea caustica*) in the Anacardiaceae family is easily the most dominant tree or shrub in the matorral (Steward and Webber, 1981) (Fig. 6.08). It grows into a tree when undisturbed, but more frequently it is found as a shrub. Its oval evergreen leaves have a smooth or undulating border. Extensive contact with the plant, by pruning it for example, can cause severe rashes and fever (the poison oak of Chile). Litre has a broader ecologic range in Central Chile than any other single matorral shrub (Rundel, 1981). Its leaves are unusually dense with a high content of lignin and cellulose, which makes them resistant to wilting, and its deep root system can take up water long into the dry season. Litre is notable for growing enormous underground burls that sprout new growth if the tree is cut (Fig. 6.07C) or burned. In addition to its unusual drought resistance and sprouting ability, litre thrives even in areas that are heavily grazed by goats (Fuentes and Muñoz, 1995).

Quillay. Quillay or soapbark tree (*Quillaja saponaria*) has an oak-like appearance and grows to a medium height (Fig. 6.09). It is in the rose family and can be recognized by its distinctive five-sectioned seed capsule. The leaves are elliptical with toothed edges. Although the tree is evergreen, many leaves are lost during the summer dry season. The inner bark foams like soap when mixed with water and is used for laundering and as a source of shampoo. Like litre, quillay tolerates a wide range of conditions: dry, rocky soil, sunny or shady slopes, and snow and wind at elevations up to 6,500 feet (2,000 m). The tallest trees grow in moist canyons, but a long taproot helps to assure the survival of the smaller trees on dry slopes.

Espino. In the Central Valley of Chile, much of the land is used for irrigated farming. Surrounding, non-irrigated areas have long been heavily grazed and used as a source of

FIG. 6.08. *Litre* (Lithraea caustica) *has simple, alternate leaves with marginal veins. The small yellow flowers have five petals, with ten stamens in the male flowers. Like poison oak, its California relative, it can cause a rash after contact; both belong to the Anacardiaceae family. (Drawing by Ràfols, 1977, reproduced with permission of the Institute of Ecology.)*

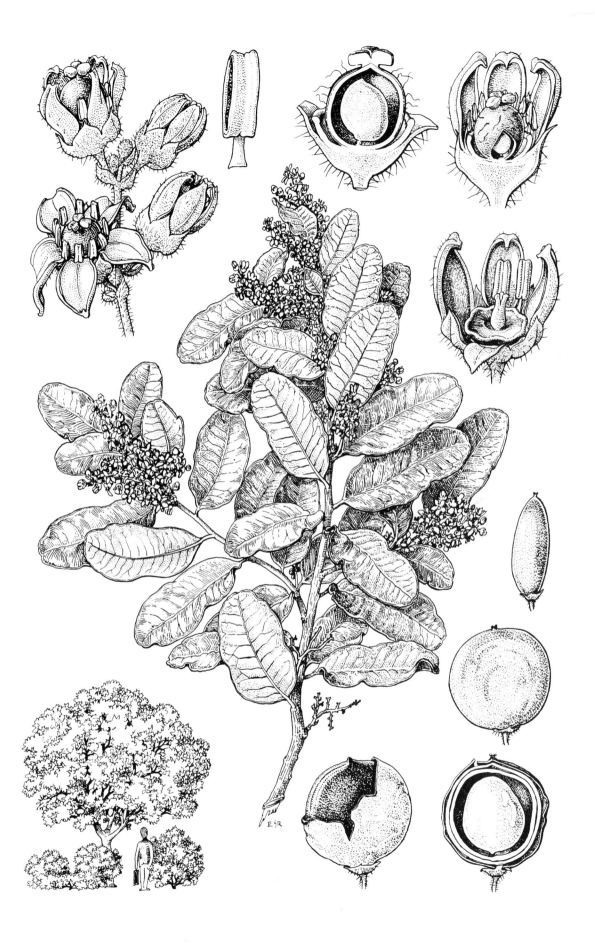

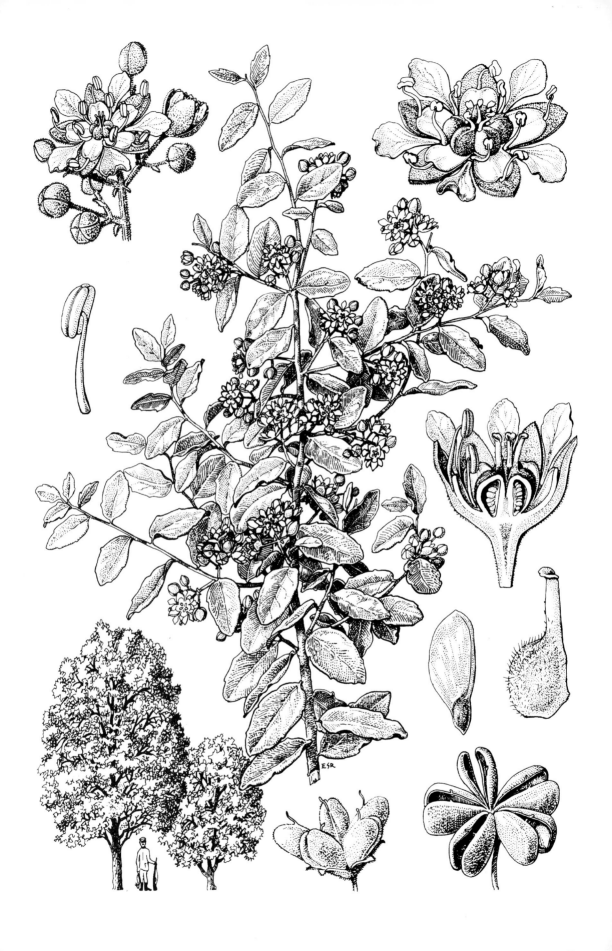

wood for fuel. This has resulted in the displacement of matorral by a sparser savanna landscape called *espinal*. Espinal is named after the thorny espino (*Acacia caven*). Dark, rounded clumps of espino form a distinctive pattern against the low-growing exotic grasses that originated in the Mediterranean Basin (Fig. 6.07D). Like many species of *Acacia* in South Africa and Australia (Fig. 6.07A), espino bursts into bright yellow bloom in early spring. *Acacia caven* is common in matorral and even grows in sclerophyll forests. Another common species in espinal is the spiny carob tree (*Prosopis chilensis*), which like espino, is in the pea family, the Fabaceae. Much of the espinal has evolved from exploited and degraded matorral. Other areas may not have changed much since before European settlement (Rundel and Weisser, 1975). Loss of a matorral vegetation cover in the Central Valley has resulted in erosion and flooding, as in other mediterranean climate regions. To control erosion, for use as windbreaks, and to supply lumber, exotic eucalyptus and Monterey pine have been planted extensively. These changes along with the spread of espinal have drastically altered the landscape.

Espinal and matorral habitats were first colonized by grasses which were introduced with winter-sown grain after settlement by Spanish colonists in the mid-16th century. This occurred much earlier than the introduction of Mediterranean Basin grasses into California in the 18th and 19th centuries. Native grasslands, which can still be found in California and in the Mediterranean Basin, are conspicuously absent in Chile.

More Woody Plants

Other Evergreen Sclerophyll Trees. The peumo (*Cryptocarya alba*), a common tree in the sclerophyll forest (page 111), is also frequently encountered as a shrub or small tree in matorral. Peumo is one of several shrub or tree species that grows smaller in matorral and than in sclerophyll forests. Other common shrubs or small trees in matorral include coliguay (*Colliguaja odorifera*), in the spurge family, which favors dry, rocky sites. It has oblong, serrated leaves and elongated, red-yellow flower clusters. Coliguay bears a three-part seed capsule that opens explosively to scatter seeds away from the parent plant. It is a particularly vigorous resprouter after cutting or burning (Fuentes and Muñoz, 1995). Olivillo (*Kageneckia angustifolia*) is a spiny shrub or small tree in the rose family. It grows on sunny slopes of the Andean foothills and can be recognized by its distinctive, five-pointed, star-shaped seed capsule.

Drought-Deciduous Shrubs. Some matorral plants, particularly in the drier, northern part of Central Chile and along the coast (page 105), lose almost all of their leaves during the summer dry season. These include several tall, spiny shrubs with green, photosynthetic

FIG. 6.09. *Quillay (Quillaja saponaria) is another common sclerophyll tree of the matorral. Its species name comes from its saponin-rich gray bark, which is still used to make shampoo. The small, star-shaped, yellow-green flowers have five sepals, five petals, and ten stamens. The seed capsule is also star-shaped with five chambers, which scatter abundant, easily germinating seeds. (Drawing by Ràfols, 1977, reproduced with permission of the Institute of Ecology.)*

stems, such as tevo (*Trevoa trinervis*) (Fig. 6.11), which has small yellow-green flowers. Other drought-deciduous shrubs are palhuén (*Adesmia microphylla*), a member of the pea family and the wild coastal fuchsia, or palo de yagua (*Fuchsia lycioides*).

A

B

Herbaceous Species

Herbaceous annuals and perennials may account for more than 40 percent of the ground surface area in matorral. One of the prominent genera with an exclusively South American distribution is *Calceolaria* (Fig. 6.12C), which is in the foxglove family (Scrophulariaceae). There are 86 species, the vast majority of which are endemic to Chile. Many herbaceous species, particularly annuals, are in the same genera as are those found in California. Sixty-six percent of the native Chilean genera of annuals are also found in California (Arroyo et al., 1995) and 41 percent of native California genera of annuals are also found in Chile. This is in sharp contrast to trees, which have only two shared genera, willow (*Salix*) and mesquite (*Prosopis*). There are fewer species of annuals in Chile than in California, where a higher fire frequency appears to favor annuals over perennials (Arroyo et al., 1995).

Geophytes. Among the best-known genera of geophytes is *Alstroemeria*, usually classified in the family Alstroemeriaceae. Flowers are azalea-shaped and are primarily deep yellow, orange (Fig 6.10A), red, white, and pink, often speckled with darker colors. They are borne in clusters on leafy stems that grow to three feet (1 m) tall. In the Amaryllidaceae (the daffodil family) are various species of the genera *Rhodophiala* and *Phycella*, commonly known as añañuca, that bear clusters of tubular or star-shaped flowers on a fleshy stalk up to 1.5 feet (0.6 m) tall (Fig. 6.12A).

Vines. Vines are probably a more prominent feature of the mediterranean vegetation

FIG. 6.10 (above). *Herbaceous plants. A. Alstroemeria spp. are common and varied bulbs in the mediterranean climate region of Chile. B. Pangue (Gunnera tinctoria) producing new leaves in early spring.*

FIG. 6.11 (opposite). *Tevo (Trevoa trinervis) is a spiny, drought-deciduous shrub of matorral. (Drawing by Ràfols, 1977, reproduced with permission of the Institute of Ecology.)*

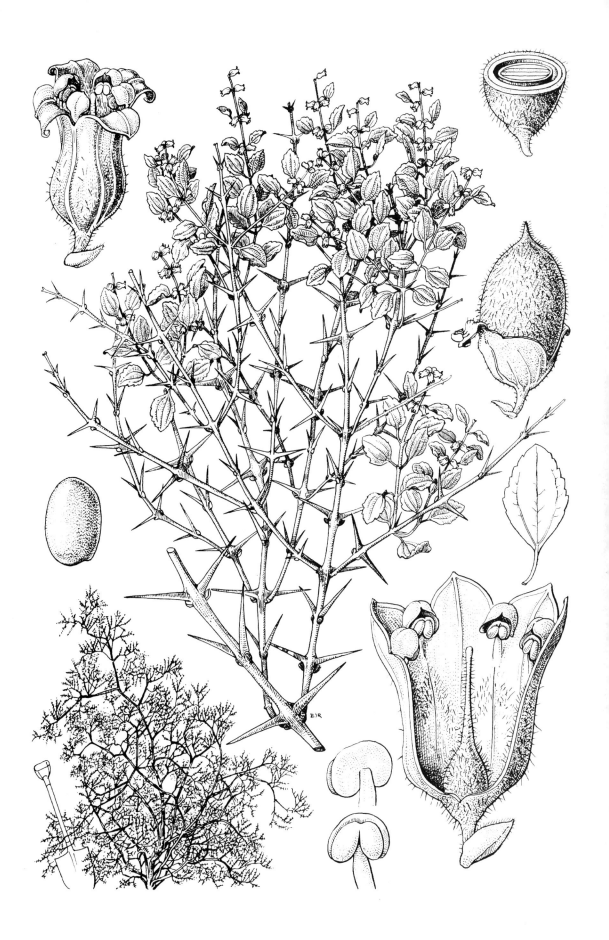

A

B

C

FIG. 6.12. Herbaceous plants. A. Añañuca (Rhodophiala advena), a bulb in the daffodil family. (Reproduced with permission of Adrianna Hoffmann.) B. Pangue (Gunnera tinctoria) is a moisture-loving plant with enormous leaves on spiky stems. These plants were photographed in mid spring near the entrance of Rio Clarillo National Reserve just south of Santiago. C. This member of the genus Calceolaria has yellow, cup-shaped flowers in mid spring and is in the foxglove family (Scrophulariaceae).

of Chile than in any of the other regions. This is thought to be related to the presence of tropical rainforest in Central Chile between about 25 and five million years ago (Arroyo et al., 1995). Many trailing and twining vines, such as the colorful relicario (*Tropaeolum tricolor*) (Fig. 6.13A) belong to the same genus as *T. majus*, the garden nasturtium. Salcilla (*Bomarea salcilla*) has attractive red flowers (Fig. 6.13B). The best known vine is the copihue (*Lapageria rosea*), Chile's national flower (Fig. 6.13C). It is most common at the southern edges of the mediterranean climate region between Concepción and Temuco. This striking deep red, bell-shaped flower is hummingbird-pollinated and blooms from spring to autumn on evergreen vines that climb up shrubs and trees.

A

Coastal Matorral

The largest stretch of coastal vegetation in the mediterranean climate belt of Chile extends from Valparaiso north to La Serena. This area becomes progressively drier toward the north, with a gradual increase in the number of succulent plants. The entire area has many low shrubs with soft foliage, analogous to the coastal sage scrub of Southern California. These include a bushy coastal daisy (*Bahia ambrosioides*). Two of the previously mentioned drought-deciduous shrubs, palhuén (*Adesmia microphylla*) and the wild coastal fuchsia, or palo de yagua (*Fuchsia lycioides*), are particularly prominent along the coast. Wild coastal fuchsia accounts for almost 25 percent of the ground cover in many places (Rundel, 1981). Its red-pink, tubular flowers are pollinated by hummingbirds and are in bloom for much of the year.

B

Tall, columnar bromeliads (*Puya*) and cacti (*Trichocereus*) are a striking and characteristic feature of the landscape of Central Chile. Both plants favor sunny, north-facing slopes of the coast ranges and Andean foothills. Both genera are poorly adapted to fire. The greater prominence of cacti in Chile than in California may be related to a lower fire frequency in Chile (Arroyo et al., 1995).

FIG. 6.13. *Vines. A. Relicario* (Tropaeolum tricolor)*, a species of what is commonly known as nasturtium. B. Salcilla* (Bomarea salcilla)*, a vine with clusters of bright red flowers. C. Copihue* (Lapageria rosea)*, the national flower of Chile.*

C

FIRE

Fires in Central Chile are relatively infrequent, except for those caused by man (Fuentes et al. 1994). The relative rarity of natural fires can be explained, at least in part, by less dense vegetation, milder summers, and an absence of summer lightning. Nevertheless, stump sprouting, which is an effective response to fire, is a common feature of matorral shrubs and trees. Sprouting is particularly vigorous in the deep-rooted species, litre (*Lithraea caustica*) (Fig. 6.08) and quillay (*Quillaja saponaria*). It is possible that stump sprouting evolved in response to pressures other than fire, such as drought and herbivory (Rundel, 1981). Alternatively, it is possible that volcanic eruptions could have resulted in periodic fires, even in the absence of summer lightning (Fuentes et al., 1994). The presence today of over 20 active volcanoes in Central Chile lends credence to this possibility.

Fire-stimulated seed germination is less common than in other mediterranean climate regions. Among the few species that exhibit this characteristic are the vine quilo (*Muehlenbeckia hastulata*) and the spiny, drought-deciduous shrub tevo (*Trevoa trinervis*).

Bromeliads. Bromeliads (the pineapple family, Bromeliaceae) are found primarily in the tropical Americas, usually growing as epiphytes on trees in the rain forest. The only mediterranean climate region in which this family is native is Central Chile. The genus *Puya* is primarily Chilean. It is terrestrial and has a fully developed root system. *Puya chilensis* has a basal rosette of stiff, spiny leaves, from the center of which a flower stalk rises to 12 feet (4 m) bearing abundant, bright yellow flowers in spring (Fig. 6.14B, D). *P. berteroniana* grows to a similar height and has very unusual blue-black flowers with bright orange stamens (Fig. 6.14A, E). The common name for both is chagual and their tall flowering spikes punctuate the landscape. Another puya, the smaller chagual chico (*P. venusta*), has a shoulder-height

A

FIG. 6.14. *Terrestrial bromeliads are a common and distinctive feature on sunny slopes of coastal and montane matorral. A. Chagual (*Puya berteroniana*) is growing on a north-facing slope in the Coast Range in La Campana National Park. Below in the park's ravine and on the opposite slope, there is a sharp change to sclerophyll forest. A closeup of chagual opposite (E) shows its unusual blue-black flowers, which highlight its bright yellow stigma and orange stamens. B, D. Puya chilensis (also commonly called chagual) sprouts tall, asparagus-like flower stalks with yellow blossoms in the spring. C. The bromeliad, Fascicularia bicolor, shows striking red leaf color before it comes into its blooming period with a relatively inconspicuous flower.*

C

B

D

E

A

B

FIG. 6.15. *Cacti and palms. A. Chilean palms (Jubaea chilensis) growing with tall cactus in La Campana National Park in the Coast Range near Santiago. B. Large, solitary, cream-colored flower of the tall cactus (Trichocereus chilensis). C. Tall cactus is common on sunny, north-facing slopes (opposite). Cacti, found almost exclusively in the western hemisphere, are a more prominent part of the mediterranean climate flora of Chile than they are in California.*

flower stalk with bright red, tubular blooms in the spring.

A second Chilean genus of bromeliads is *Fascicularia*, with five species. The genus is primarily terrestrial but can also grow as an epiphyte on the branches of trees (Wilcox, 1996). It is stemless or short-stemmed with a dense, flat rosette in the center of which the flower is embedded. Chupalla (*Fascicularia bicolor*) grows on rocky, ocean-facing cliffs, in the coast ranges, and in the moister rainforest to the south (Fig. 6.14C). Its leaves are most conspicuous when their innermost portions turn a bright red during the one to two months before its blue-petalled flowers bloom.

Cacti. The cactus family (Cactaceae) is restricted almost entirely to North and South America. Cacti grow primarily in warmer, semi-arid climates. They are spiny succulent plants that are nearly or entirely leafless and that photosynthesize in their green trunks. Many Cactaceae have a superficial resemblance to the unrelated, spiny, succulent members of the genus *Euphorbia*, which are common in the Karoo, at the dry end of the mediterranean climate zone of the Western Cape in South Africa.

Cacti are far more prominent and diverse in Central Chile than they are in corresponding parts of California that have a similar climate. The common cactus, *Trichocereus chilensis*, is recognizable from a distance by its tall, cylindrical form, rising to heights of 23 feet (7 m) (Fig. 6.15C). Its solitary flowers are cream colored (Fig. 6.15B), and its fruit is edible. The coastal cactus (*Trichocereus litoralis*) is cylindrical and about shoulder height. Its large, solitary, cream-colored flowers bloom in the spring, always emerging on the sun-facing, north side of the plant. Quisquito, or pink

C

cactus (*Neoporteria subgibbosa*) is cylindrical and sometimes branched and grows to three feet (1 m) tall. Its pink flowers also bloom in the spring and are grouped near the apex. Cacti become progressively more abundant as the climate becomes drier from Valparaiso north toward La Serena.

Woodland and Forest

Before the Spanish settlement of Central Chile, the Central Valley, the coast ranges, and the Andean foothills were more wooded. Nevertheless, within ten years of the founding of Santiago in 1541, it was necessary to establish rules to control the cutting of nearby foothill trees for fuel, construction, and supports for mining tunnels (Hoffmann and Fuentes, 1988).

Sclerophyll Woodland and Forest

The scant sclerophyll forest now remaining is located mainly in the coast ranges, between Valparaiso and Concepción, and in the Andean foothills at about the same latitudes

FIG. 6.16. *Southern beech* (Nothofagus)*. Roble* (N. obliqua) *forest in autumn color in the foothills of the Andes south of Santiago.* (Photograph by Adrianna Hoffmann, reproduced with her permission.)

(Fig. 6.01) (Wilcox, 1996). One of the best places to see the woodland of the coast ranges in relatively undisturbed form is in La Campana National Park (Fuentes and Hoffmann, 1988), less than a two-hour drive from the Santiago area and about one hour from Valparaiso. Evergreen sclerophyll forest grows in the moister canyons and south-facing slopes of the coast range, as illustrated on Figs. 6.06 and 6.07B.

Peumo. Peumo (*Cryptocarya alba*) is one of the most common broadleaf evergreen trees of Central Chile (Steward and Webber, 1981) (Fig. 6.17). It is found in the coast ranges as well as in the Andean foothills. The tree has dense foliage with smooth, bright green, oval leaves, which have a pleasant fragrance when bruised. It belongs to the laurel family (Lauraceae), which includes the California bay (*Umbellularia californica*) and the

FIG. 6.17. Peumo (Cryptocarya alba) *can grow as a tall, dense tree, despite being relatively drought-tolerant. It has small, inconspicuous, yellow-green flowers, but the fruit is intense red and the size and shape of an olive.* (Drawing by Ràfols, 1977, reproduced with permission of the Institute of Ecology.)

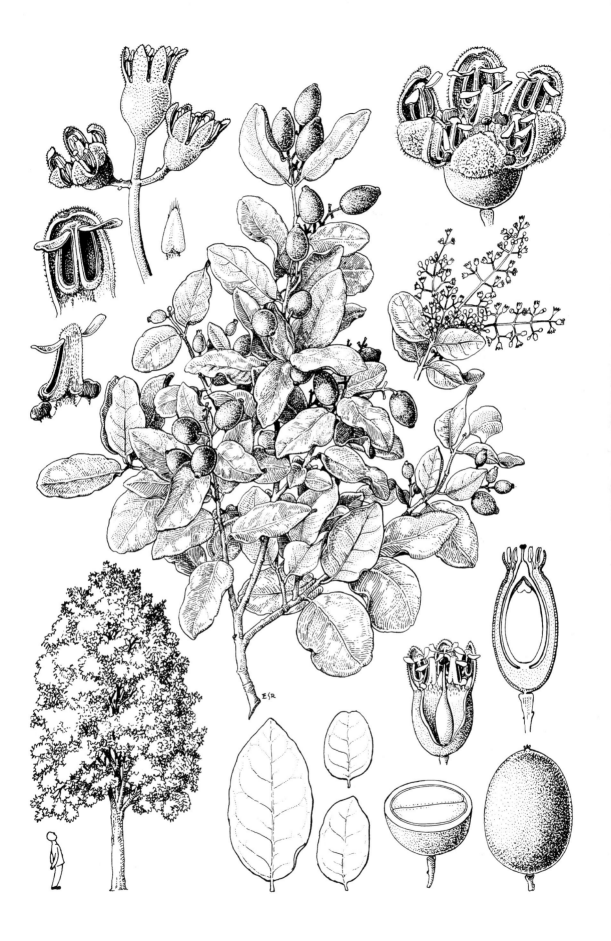

bay laurel (*Laurus nobilis*) from the Mediterranean Basin, both of which are also known for their pungent leaves.

The peumo can reach a height of 65 feet (20 m) in the moister parts of its range. Although its yellow-green flowers are inconspicuous, its olive-size, bright red fruit is more visible. In the drier, matorral plant community of the mediterranean climate area, peumo is found mainly on shady, south-facing slopes and in canyons, where it can form almost pure groves (Hoffmann, 1982). Farther south, where there is a longer rainy season, the tree is found in drier, more sun-exposed locations. Its root system is relatively shallow, and its leaves are not as drought adapted as those of many matorral shrubs and trees. However, there is a seasonal variability or dimorphism in the size of leaves (page 33), which is effective as an adaptation to drought (Steward and Webber, 1981). Late summer leaves resist water loss by growing to only a fifth of the size of leaves produced in the winter-to-spring growing season.

Boldo. Boldo (*Peumus boldus*), in the mainly southern hemisphere Monimiaceae family, is another common evergreen tree. It has edible fruit, a bark that has been used for tanning leather, and wood that makes excellent charcoal. Boldo reaches a height of 20 feet (6 m) and resembles an evergreen oak from the distance. It grows on relatively dry and rocky soil through much of Central Chile.

Mayten. Mayten (*Maytenus boaria*), in the Celastraceae family, is an attractive evergreen shade tree that can have either a weeping, willow-like habit or a more upright form. It favors moist places near streams. The maitén is widely used as a street tree in California.

Chilean Palm. Chilean palm (*Jubaea chilensis*) grows to 50 feet (15 m) tall and has a distinctively thick trunk (Fig. 6.15A). It was once very abundant in the valleys and lower slopes of the coast ranges, but most of the palms were cut for their abundant sweet sap, which continues to flow for months from the periodically severed end of a chopped down, horizontal trunk. The sap was concentrated like maple syrup to make *miel de palma*, or palm honey. In the 17th and 18th centuries, this was the major source of sugar for the entire country. On the northern side of La Campana National Park is the Ocoa Valley, which has one of the few remaining large stands of Chilean palm, with thousands of trees dominating a mountainous landscape.

Associated Species. Sclerophyll woodland and forest also extend into the Andean foothills to an elevation of about 4,600 feet (1,400 m). Near Santiago, Río Clarillo National Reserve includes such woods in the canyon of a branch of the Maipo River. At higher elevations are shrubs and alpine plants with scattered mountain cypress, *Austrocedrus chilensis*. Among the flowering herbaceous plants are *Tropaeolum, Alstroemeria, Hippeastrum*, and *Calceolaria*.

A

B

Moister Forests

With increasing rainfall and a shorter dry season, other groups of broad-leaved evergreen trees become dominant. The greatest diversity of forest species in Chile is between the cities of Talca and Valdivia (Villagran, 1995).

Nothofagus Forest

Various species of southern beech (*Nothofagus*) predominate in the southern part of the mediterranean climate area of Central Chile and south of the Río Bío Bío, near Concepción (Veblen et al., 1996). Southern beech is in the same family as oaks and sweet chestnuts, the Fagaceae.

Roble. The deciduous roble (*Nothofagus obliqua*) grows further north than any other species of southern beech and is represented by sizable stands at higher elevations on the shady south- and southeast-facing slopes of the coast range in La Campana National Park (Rundel and Weisser, 1975). South of Concepción, where there is a longer rainy season, roble is one of the most common trees in the Central Valley and on the lower mountain slopes. The trees are deciduous and turn attractive yellows and reds in the fall (Fig. 6.16). The roble can reach a height of 130 feet (40 m). It is a sprouter that regrows in slender stems from cut or burned stumps.

FIG. 6.18. *Monkey puzzle tree* (Araucaria araucana). *A. The tall gray trunks of mature* A. araucana *trees have a deeply incised cobblestone pattern.* B. Young Christmas tree-shaped and old, tall, umbrella-shaped A. araucana *trees in Nahuelbuta National Park.*

Coihue. Coihue (*N. dombeyii*) is one of three evergreen species of *Nothofagus* in Chile. It is probably the most abundant native forest tree between Concepción and Puerto Montt. In the springtime the dark green forests of coihue, in Nahuelbuta National Park for example, form a sharp contrast against the fresh, pale green of the deciduous *Nothofagus* species. Coihue is the tallest broadleaf tree in Chile, reaching a height of 165 feet (50 m).

Other Trees. Among other moisture-loving trees are patagua (*Crinodendron patagua*) in the Elaeocarpaceae

A

B

FIG. 6.19. *Flowering trees and shrubs of the moister, southern part of central Chile. A. Canelo (Drimys winteri) is an evergreen tree that is sacred to the Mapuche Indians. B. Chilean fire bush (Embothrium coccineum) is a South American representative of the protea family. C. Pelú or Sophora tetraptera var. microphylla is of special interest because the same species is also found across the Pacific Ocean in New Zealand. D. Fuchsia magellanica is an abundant shrub with red and violet tubular flowers. E. Tupa (Lobelia exelsa).*

D

C

E

family, which has oblong serrated leaves and small, white bell-shaped flowers in late spring. Arrayán (*Luma apiculata*), in the myrtle family, has a distinctive, smooth, rust-colored, twisted trunk, reminiscent of madrone in California. Canelo (*Drimys winteri*) in the Winteraceae family is a sacred tree used in the healing and agricultural fertility ceremonies of the Mapuche Indians. Its bark was exported in the 19th century to prevent scurvy. It has clusters of white flowers borne on stems that turn red when the tree is in bloom (Fig. 6.19A).

Understory Plants

Notro. Notro or Chilean fire bush (*Embothrium coccineum*) is one of the prominent flowering shrubs in the moister, southern part of Central Chile (Fig. 6.19B). This common species is a South American representative of the protea family, which has a much greater diversity of genera and species in the Western Cape and in Australia. A few species are also found in New Zealand. This southern hemisphere distribution of the family supports strong evidence for the past existence of a single southern hemisphere continent of Gondwana (page 26).

Pelú. Pelú or *Sophora tetraptera var. microphylla* is a shrub in the pea family (Fabaceae) with attractive yellow flowers (Fig. 6.19C). It is of special interest because the same species is also found in New Zealand, where it is the national flower known as kohai, the Maori word for yellow. Although this distribution suggest an origin in the ancient continent of Gondwana, it could have spread across the Pacific Ocean by seed. Seed capsules stay afloat over long distances, bearing viable seeds (Sykes and Godley, 1968).

Fuchsia. A common species of *Fuchsia* in the evening primrose family, is *F. magellanica* (Fig. 6.19D), which has a remarkably wide distribution, extending from well north of semi-arid La Serena all the way to Tierra del Fuego, in the extreme south. It is an abundant shrub in the wetter parts of Central Chile. Its red and violet tubular flowers bloom in the winter half of the year and are pollinated by hummingbirds.

Tupa. Tupa or *Lobelia exelsa* is an evergreen perennial with bright red, tubular flowers (Fig. 6.19E). It contains a poisonous alkaloid, accounting for one of its common names, tabaco del diablo or devil's tobacco.

Pangue. Among the common herbaceous understory plants along streamsides and other moist locations in *Nothofagus* forest is pangue (*Gunnera tinctoria*), with enormous annually-produced leaves on spiky stalks (Figs. 6.10B, 6.12B). The plant is most abundant in the moister, southern part of Central Chile, where young stalks are peeled and eaten with salt as a crisp, refreshing snack.

Bamboo. One of the most common understory plants in *Nothofagus* forest is the bamboo, *Chusquea quila*. It forms an impenetrable scrub called quilantal. Together with blackberry (*Rubus constrictus*), it invades areas where forest has been cut, making it difficult for the original forest community to reestablish itself (Ramírez et al., 1989).

Araucaria Forest

At the moist, southern end of the mediterranean climate area are remainders of once extensive forests of the araucaria or monkey puzzle tree (*Araucaria araucana*) (Jung et al., 1992). Araucaria forests are found in national parks in the Andean foothills (Conguillío National Park) and in the coast ranges (Nahuelbuta National Park) (Fig. 6.18B) at about 38° latitude. The genus *Araucaria* is a primitive conifer that is also found in southern Brazil, in eastern Australia, and on Norfolk Island off the eastern Australian coast. Mature *A. araucana* are as tall as 165 feet (50 m), deep green, and umbrella-shaped on thick gray trunks with a deeply incised cobblestone pattern (Fig. 6.18A). Monkey puzzle trees are also commonly known as paraguas, or umbrellas, in Spanish and as pehuén in the language of the Mapuche native peoples.

Monkey puzzle trees occur in pure stands, often on hilltops where, unlike most other Chilean forests, they are easy to walk through. Some stands are accompanied by smaller, deciduous species of southern beech (*Nothofagus*), making the evergreen monkey puzzle trees more prominent in winter and early spring. Huge, pineapple-size cones are borne in mid-summer on the top of female trees. Each cone has as many as 200 seeds that are tasty and nutritious and were once a staple food of the Mapuche Indians. Male trees produce enormous amounts of pollen, referred to as sulfur rain (Wilcox, 1996). The broad tops of the trees are popular nesting sites for large birds, such as eagles, hawks, and owls.

Alerce Forest

Forests of alerce or Fitzroy cypress (*Fitzroya cupressoides*) grow in the cool, moist slopes beyond the southern end of the mediterranean climate region of Chile. They are nevertheless worth mention because, together with the coihue, they are the largest trees in Chile, growing to 165 feet (50 m) in height and more than ten feet (3 m) in trunk diameter. They have been heavily logged because their wood is attractive and extremely resistant to decay. The alerce reaches ages of 4,000 years. During Charles Darwin's travels through Chile in 1834 and 1835, he wrote about the extensive alerce forests of the island of Chiloe. Darwin's friend, the botanist Joseph Hooker, later named the alerce *Fitzroya cupressoides* after Fitzroy, the captain of Darwin's ship, the Beagle. Alerce grows at a slow rate, so that second-growth forests have not formed as well as they have for the coast redwood in California. Some of the few remaining large stands are now protected in national parks and nature reserves.

PLANTS FOUND IN MEDITERRANEAN CLIMATE AREAS OF CHILE

Acacia caven	espino	Fabaceae	spiny shrub/tree
Alstroemeria spp.	alstromeria	Alstroemeriaceae	bulbous
Araucaria araucana	monkey puzzle tree	Araucariaceae	evergreen conifer tree
Austrocedrus chilensis	ciprés	Cupressaceae	evergreen conifer tree
Bomarea salsilla	salsilla	Alstroemeriaceae	herbaceous vine or herb
Calceolaria spp.	calceolaria	Scrophulariaceae	shrub
Cryptocarya alba	peumo	Lauraceae	evergreen tree
Drimys winteri	canelo	Winteraceae	evergreen tree
Embothrium coccineum	notro	Proteaceae	shrub
Fitzroya cupressoides	alerce, Fitzroy cypress	Cupressaceae	evergreen tree
Flourencia thurifera	maravilla del campo	Asteraceae	deciduous shrub
Fuchsia lycioides	palo de yegua	Onagraceae	shrub
Hippeastrum spp.	añañuca	Amaryllidaceae	bulbous
Gunnera tinctoria	pangue	Gunneraceae	herbaceous perennial
Jubaea chilensis	palma chilena	Arecaceae	palm tree
Kageneckia angustifolia	olivillo	Rosaceae	evergreen shrub/tree
Lapageria rosea	copihue	Liliaceae	woody vine
Lithraea caustica	litre	Anacardiaceae	evergreen shrub/tree
Maytenus boaria	mayten	Celastraceae	evergreen tree
Myrceugenia obtusa	arrayan	Myrtaceae	evergreen shrub/tree
Nothofagus dombeyi	coihue	Fagaceae	evergreen tree
Nothofagus obliqua	roble, southern beech	Fagaceae	deciduous tree
Peumus boldus	boldo	Monimiaceae	evergreen shrub/tree
Puya chilensis	chagual	Bromeliaceae	evergreen perennial
Quillaja saponaria	quillay	Rosaceae	evergreen shrub/tree
Trevoa trinervis	tevo	Ramnaceae	spiny deciduous shrub
Trichocereus chilensis	tall cactus	Cactaceae	cactus
Tropaeolum spp.	nasturtium	Tropaeolaceae	herbaceous vine

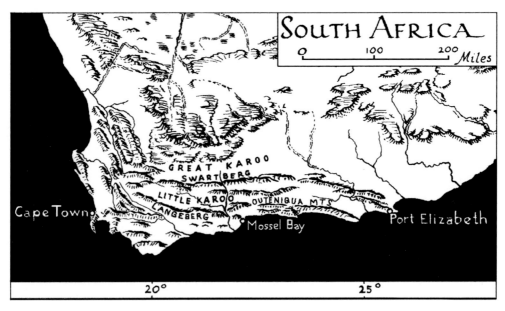

FIG. 7.01. *Western Cape region, with an orientation map below, showing the position in South Africa and the African continent. (Adapted from a map by Erwin Raisz with permission of John Wiley & Sons.)*

7.

WESTERN CAPE, SOUTH AFRICA

Landscape and Climate

The mediterranean climate region of South Africa occupies a coastal strip of land along the extreme southwestern tip of the continent (Fig. 7.01). Formerly, this area was part of Cape Province, which accounted for over half of South Africa. Recently, Cape Province was divided into three smaller provinces, Western Cape, Eastern Cape, and Northern Cape. The mediterranean climate area lies entirely in Western Cape, and is centered at its largest city, Cape Town.

Cape Town has a spectacular harbor setting with Table Mountain as a backdrop. It has a population of about two million in its metropolitan area and lies near the Cape of Good Hope, where the Atlantic and Indian oceans meet. The entire mediterranean climate area is a crescent-shaped coastal strip that comprises most of the Cape Floral Kingdom (page 20). One arm of the crescent points north and the other points east from Cape Town. The northern arm faces the Atlantic coast to the west for about 150 miles (250 km). East of Cape Town and the Cape of Good Hope, the coastline makes a right angle turn to face the Indian Ocean to the south. This part of the mediterranean climate crescent lies in the southernmost part of the African continent. It extends for about 200 miles (320 km) along the Indian Ocean coast to Mossel Bay. Along its entire length, the mediterranean climate reaches inland for about 75 miles (125 km).

The Atlantic coastal area to the north of Cape Town is largely flat and is known as the West Coast. Further inland rises a series of north-south mountain ranges, including the occasionally snow-covered Cedarberg Range. East of Cape Town lie the Hottentots Holland Mountains, which plunge steeply into the Indian Ocean near Cape Hangklip on the southern coast. The Hottentots Holland region, about a one-hour drive from Cape Town, has the greatest diversity of plant species in South Africa. Further east, two to three linear mountain ranges take an east-west direction, running parallel to the Indian Ocean coast. These include the Langeberg, Swartberg, and Outeniqua Mountains, whose highest peaks reach about 7,600 feet (2,300 meters). The long intervening valleys are largely agricultural and are known for their excellent wine and fruit.

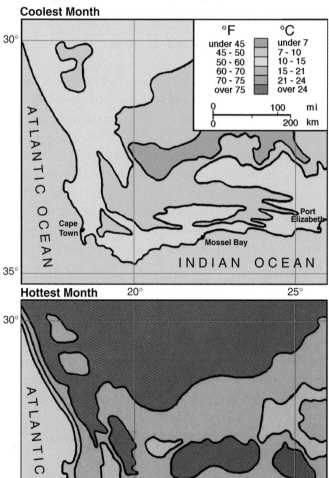

Annual Precipitation: Western Cape, South Africa

inches	cm
under 10	under 25
10 - 25	25 - 63
25 - 40	63 - 100
over 40	over 100

0 ..... 100 m
0 ..... 200 km

ATLANTIC OCEAN

Cape Town

Port Elizabeth

Mossel Bay

INDIAN OCEAN

30°
35°

Average Temperature: Western Cape, South Africa

Coolest Month

°F	°C
under 45	under 7
45 - 50	7 - 10
50 - 60	10 - 15
60 - 70	15 - 21
70 - 75	21 - 24
over 75	over 24

0 ..... 100 mi
0 ..... 200 km

ATLANTIC OCEAN

Cape Town

Port Elizabeth

Mossel Bay

INDIAN OCEAN

30°
35°

Hottest Month

20°   25°

ATLANTIC OCEAN

Cape Town

Port Elizabeth

Mossel Bay

INDIAN OCEAN

30°
35°

20°   25°

Rainfall

Rainfall is greatest on the coastal side of the mountain ranges (Fig. 7.02). In contrast, the intervening valleys that lie in the rain shadows of the mountains have less than half as much precipitation (McMahon and Fraser, 1988). In general, the climate becomes progressively drier inland and toward the north.

Winter rainfall comes in heavy storms brought by westerly winds, mainly in May, June, and July. Winter also brings occasional hot, desiccating, and gusty "berg" winds that come from the interior. During the three summer months, Table Mountain near Cape Town has a "tablecloth" of mist that adds as much as 20 inches (50 cm) of moisture per year on the mountaintop. Although there is little summer rainfall near Cape Town, there are increasing amounts as one travels east toward Mossel Bay. Lightning storms

FIG. 7.02. *Annual precipitation, Western Cape (top). Areas of highest rainfall are in the mountains and along the coastal exposures of the southern coast.* (Based on *Atlas of the Union of South Africa, 1960.*)

FIG. 7.03. *Average temperature of the coolest and hottest months, July and January, respectively.* (Based on *Atlas of the Union of South Africa, 1960.*)

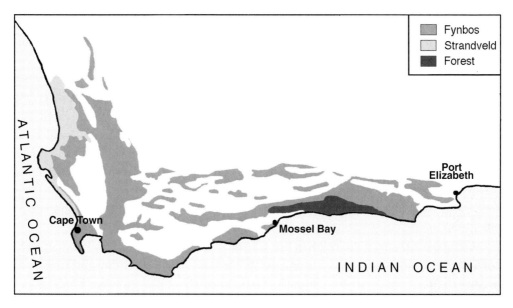

Fynbos
Strandveld
Forest

ATLANTIC OCEAN

Port Elizabeth

Cape Town

Mossel Bay

INDIAN OCEAN

FIG. 7.04. *Mediterranean plant communities in Cape Province, South Africa. (Based on Bond and Goldblatt, 1984; Acocks, 1988.)*

are common and often start brush fires, which are most frequent in summer and early fall, but can occur at any time of the year.

Temperature

Temperatures are mild with a strong maritime influence near the coast, bringing moderate summers and mild winters (Fig. 7.03). The western Atlantic coast from the Cape of Good Hope toward the north is cooled by the cold Benguela current. In the summer, persistent and often strong southeast winds called the "Cape Doctor" cool coastal areas, which consequently have less than 18°F (10°C) difference between average summer and winter temperatures. The southern coast, east of the Cape of Good Hope, is under the influence of the warmer water of the Indian Ocean. Interior valleys are dry with hot summers and more likelihood of frost in winter. Although the flanking mountains are sometimes cold enough to be snow capped in winter, there is no area in which the average temperature of the coldest month is near or below the freezing point.

PLANT COMMUNITIES

Fynbos, an Africaans term meaning fine bush, is the most extensive and varied form of native vegetation in the mediterranean climate crescent of the Western Cape (Fig. 7.04). It is dominated by evergreen shrubs of varying heights and includes over 80 percent of the plant species in the entire mediterranean climate region. Its peak bloom is in the spring months of September and October. But plant growth and flowering in fynbos extend through more of the year than in shrublands of the other mediterranean climate areas (Orshan et al., 1989; Keeley, 1992). This pattern is attributed to the subtropical and tropical origins of much of the flora. An advantage of a spread-out flowering season is that it decreases competition for pollinators.

Although the mediterranean climate area along the southern coast extends only from Cape Town to Mossel Bay, fynbos and the Cape Floral Kingdom both continue further east into the area of year-round rainfall between Mossel Bay and Port Elizabeth (Fig. 7.04). The fynbos here has an even more year-round pattern of flowers in bloom. The Outeniqua and Tsitsikamma Mountains, along the Indian Ocean coast east of Mossel Bay, have South Africa's largest remaining forests, where trees such as yellowwood (*Podocarpus* spp.) and stinkwood (*Ocotea bullata*) occur. In the mediterranean climate region between Cape Town and Mossel Bay, small patches of forest are found only along shaded streams and in mountain clefts.

Along the Atlantic coast north of Cape Town, a form of vegetation called strandveld or fynbos/thicket mosaic grows on sandy soil. In the same region, there are also large areas in which fynbos is predominant. Strandveld is dominated by lower growing shrubs, succulents, bulbs, annuals, and reed-like restios (Restionaceae). In the late winter and early

NUTRIENT-DEFICIENT SOIL

Some similarities between South Africa's fynbos and Australia's kwongan can be attributed to nutrient-poor soils (Keeley, 1992). Neither of these plant communities is as dense as chaparral or maquis. The more open growth pattern typically found on nutrient-poor soils favors a diversity among low-growing species that in chaparral is approached only in the two years after a wildfire. The low density of plants is readily apparent from the ease of walking in fynbos and kwongan compared with chaparral and maquis.

Drought-deciduous plants are not prominent in fynbos or kwongan, possibly because the nutrient-poor soil is incompatible with the additional nutrient requirements of growing new, replacement sets of leaves after the dry season.

For many fynbos shrubs, seed maturation takes an unusually long time, allowing a longer period for the accumulation of nutrients. Furthermore, plants in the protea family, Proteaceae, produce relatively few fruits or seeds despite their abundant flowers. In a number of species, the average fruit set is as low as six percent in the genus *Leucospermum* and nine percent in the genus *Protea*. An even lower mean of two percent fertile seed is found among a group of species in the Australian genus *Banksia*, which is also in the protea family and also tolerant of the nutrient-poor soils (Johnson, 1992). *Banksia laricina* doubles the number of follicle-bearing seeds with the use of nitrogen and phosphorus fertilizer, indicating that lack of soil nutrients is at least partly responsible for a low reproductive potential.

Production of protein-rich seeds is highly dependent on soil nutrients (Johnson, 1992; le Maitre and Midgley, 1992). On the other hand, production of abundant nectar and woody fruit is less demanding of soil nutrients since it relies more heavily on carbon that comes from the air. Carbon dioxide is the source of the carbon that is used for the produc-

spring months of August and September, strandveld is known for its abundance of wild-flowers.

Behind the southernmost coastal mountain range there is a long, narrow, semi-arid valley known as the Little Karoo (Fig. 7.01). Although this valley is largely in irrigated agriculture, the surrounding hills are known for their rich spring bloom of succulents in the genus *Dorotheanthus* and ice plants in the large complex of *Mesembryanthemum* genera, all in the Aizoaceae family. In addition there are many species of *Aloe* in the lily family (Liliaceae) and of daisies (Asteraceae). The Little Karoo was the natural habitat of the ostrich, which is now only commercially farmed there.

Two semi-arid regions of great botanical interest deserve mention although they lie just beyond the mediterranean climate area. The Great Karoo is a broad valley north of the Little Karoo. It has the richest diversity of succulent flora in the world, with many species that also extend into fynbos. The second region is Namaqualand, which is about 250 miles (420 km) north of Cape Town. Following adequate winter rains, this large, sparsely populated area is first to burst into bloom, usually in late July and August, in what is the most luxuriant wildflower display of the Western Cape.

tion of sugar by photosynthesis. Flower growth is also relatively independent of soil nutrients, allowing the production of unusually large and often nectar-rich flowers in the genus *Protea*. The abundance of nectar may help to explain why plant pollination by birds is relatively common in fynbos, as it is in kwongan of Australia.

The seeds of about 20 percent of fynbos species, including many in the genus *Leucospermum*, are dispersed by ants, which store seeds in their underground nests. The ants are attracted by a fat-rich appendage of the seed, called an elaiosome, which they can eat without killing the seed (Fig. 7.05). The depth at which the seeds are buried varies with the species of ant. When elaiosomes are experimentally removed, the seeds are still capable of sprouting, but since they are they are no longer buried by the ants in this form, they are eaten by rodents (Johnson, 1992). About 1300 fynbos species in 24 families have this mutually beneficial relationship with ants. The benefit to the ants is a calorie-rich food. The plant benefits by having its seeds protected from predators, such as rodents. In addition, the ant nest provides more nutrients and better soil drainage for the sprouting seed (Hölldobler and Wilson, 1990). Seed dispersal by ants is also very common in Australia, found occasionally in Chile, but rare in other mediterranean climate areas.

FIG. 7.05. *Elaiosomes. One-fifth of fynbos species have seeds with fat-rich appendages called elaiosomes. The appendages attract ants which store the seeds in their underground nests. The ants benefit by eating the nutritious appendage, and the plants benefit by having the still viable seeds efficiently dispersed and protected from other animal predators. (Reproduced with permission from Hölldobler and Wilson, 1990.)*

Fynbos

As does kwongan of Western Australia (page 146), fynbos has a remarkably large variety of plant species, many of which are endemic to very small areas. Although fynbos habitat is found close to the coast as well as in mountains, a greater expanse of mountain fynbos has survived because its rugged terrain and nutrient-poor soil make cultivation difficult. Coastal fynbos is threatened by extensive urbanization of attractive coastal landscapes, particularly within a one- to two-hour drive of Cape Town.

Variety in height and diversity of species characterize fynbos as well as kwongan in Southwestern Australia (le Maitre and Midgley, 1992). This contrasts with a more uniform shrub height and smaller number of dominant species found in California chaparral (Keeley, 1992). Other distinctive features that fynbos shares with kwongan are an abundance of species that retain seeds in the canopy until a fire, numerous species with seed dispersal by ants, and a rarity of fleshy fruit and bird-dispersed seeds. Both regions have many species from the family Proteaceae. However, in place an abundance of species of *Erica* in the Western Cape, heaths in the closely related family Epacridaceae are found in Australia. The prominent myrtle family, Myrtaceae, which includes species of *Eucalyptus* that dominate the Australian landscape, is not well represented in the Western Cape, except by plants imported from Australia that later proved to be invasive.

FIG. 7.06. *Fynbos with oleander-leaved protea* (Protea neriifolia) *against the background of Paarl Mountain.*

FIRE

*F*ires are even more common in fynbos than in California chaparral. It is rare to find fynbos stands that are more than 20 years old, whereas many chaparral stands survive over 20 years and quite a few are over 50. This results from the extreme flammability of dried branches, stems, and leaves of *Erica* and Restionaceae. The risk of highly destructive fires is decreased by a generally accepted policy of using prescribed burns to avoid dangerous accumulations of fuel and of allowing wildfires to burn if they do not threaten homes and other developed areas.

The survival of dominant fynbos plant species despite frequent fires is favored by the rapid rate at which they mature to a reproductive stage. Many fynbos shrubs mature in less than seven years, whereas the period of maturation of many chaparral shrubs is longer (Keeley, 1992). Fire stimulates subsequent growth of seedlings by returning nutrients from leaf litter to the soil. Between fires, leaf litter breaks down more slowly in fynbos than it does in California chaparral or Australian kwongan.

In contrast to other regions with a mediterranean climate, regrowth of shrubs after a fire in fynbos is largely from stored seeds dropped from the canopy and from seeds stored below the soil surface. Relatively few of the dominant shrubs sprout from the base after a fire. Some examples of stump-sprouting shrubs are *Protea nitida*, *P. cynaroides*, and several species of the sunshine protea, *Leucadendron*.

Fynbos is made up primarily of low-growing heath (the genus *Erica*) and rush-like Restionaceae, with members of the protea family (Proteaceae) that range from shrubs to small trees (Fig. 7.06). The iris, amaryllis, and daisy families are also well represented. In addition, there are succulents such as *Euphorbia* and *Aloe* but they are less abundant and diverse than in adjacent, drier regions of the Karoo.

Protea Family

The Proteaceae include several showy genera that are well-known to horticulturists (Matthews and Carter, 1993). The genus *Protea*, after which the family is named, dominates large areas of fynbos. Another genus in the same family, *Banksia* is correspondingly prominent in Australia (page 152). Proteaceae are also found in Central and South America and are represented in the moister parts of Central Chile by the Chilean fire bush (*Embothrium coccineum*) (page 114). Fossil pollen of Proteaceae has also been found in Antarctica, providing evidence for the presence of this family in the ancient, southern hemisphere continent of Gondwana.

The Genus Protea. The 114 species of the genus *Protea* are all found in sub-Saharan Africa. The genus probably originated in tropical East Africa, where it still has a broad distribution with species that have the simplest flower structure. Of the 82 species found in

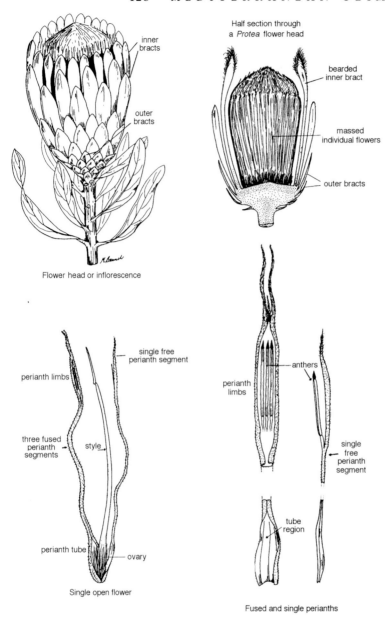

Flower head or inflorescence

Half section through
a *Protea* flower head

bearded
inner bract

massed
individual flowers

outer bracts

Single open flower

Fused and single perianths

FIG. 7.07. Protea *structure. Protea flower heads are thistle-like in structure, with bisexual flowers grouped in bowl-shaped heads that are surrounded by conspicuous, often brightly colored bracts (modified leaves at the base of a flower). Each of the many individual flowers is made up of four anthers, of which three are typically joined and one is free. There is a single, one-chambered ovary with a single style and a tiny stigma. (From Rourke and Anderson, 1982.)*

South Africa, about 69 are concentrated in the narrow coastal strip of the Western Cape, many occupying very small ranges. They are highly varied and are eloquently described as representing "an evolutionary outburst of morphological experimentation" (Rourke and Anderson, 1982). Not surprisingly, the novel and diverse proteas astounded early botanical explorers from Europe between late 17th and early 19th centuries.

Protea flower heads are large and thistle-like in appearance. The flowers are bisexual and are grouped in bowl-shaped heads that are surrounded by conspicuous, often brightly colored bracts (modified leaves at the base or around a flower) (Fig. 7.07). Each of the many individual flowers contains four anthers, of which three are typically joined and one is free. There is a single, one-chambered ovary with a single style and a tiny stigma. Based on this flower structure, Linnaeus characterized the genus *Protea* as having four husbands in bed with one wife. The tallest plants found in fynbos are generally the taller species of *Protea*, which grow well above the *Erica* and Restionaceae.

Many species of *Protea* retain their seeds in the canopy to release them only after a fire.

The success with which they reproduce depends to a large extent on the duration between the fire and autumn, when seeds germinate with the onset of the rainy season. Seeds released after winter and spring fires are largely eaten by rodents, with few survivors left to produce seedlings. The more common summer and autumn fires expose seeds to rodents for a shorter period and result in larger crops of seedlings (Johnson, 1992).

King Protea. King protea (*Protea cynaroides*) is the official floral symbol of South Africa. The species name comes from the resemblance of the flower bud to the artichoke, *Cynara scolymus*. The king protea is common in much of the fynbos region (Rourke and Anderson, 1982). It grows from sea level to elevations of about 5,000 feet (1,500 m), where it periodically becomes snow-covered in winter. The king protea reaches six feet (2 m) at maturity and grows as a single plant, rather than as large clumps. The plant has a woody lignotuber that sprouts within a few months of a fire, a survival adaptation to wildfires. Its huge flower heads are goblet- or bowl-shaped (Fig. 7.08A) and range from four to 10 inches (10 to 30 cm) in diameter. Bracts are often pale pink, but may also be crimson or bronze-red. Like *P. neriifolia* (Fig. 7.06) and *P. repens*, the king protea has numerous local varieties that differ not only in structure, color, and size, but notably in the season in which they are in bloom. The variation in flowering time is a genetic characteristic that is convenient for commercial growers, who can supply flowers throughout the year.

Sugarbush. Sugarbush (*Protea repens*) receives its popular name for its abundant, sweet nectar. Flower heads actually drip nectar when they are inverted. From as early as 1720, settlers collected the nectar, filtered it, and thickened it by boiling (Rourke and Anderson, 1982). The resulting brown or amber "Bossiestroop" was used as a sweetener and a cough syrup. It was still listed as a remedy called *Syrupus proteae* in the mid 1800s. Sugarbush was the first protea to be cultivated away from the Cape. The plant collector, Francis Masson, succeeded in establishing it in the Royal Collections at Kew Garden near London in 1774, where it flowered a few years later and started the craze for protea growing among the European aristocracy. Sugarbush, in contrast to king protea, grows in large stands averaging about six feet (2 m) tall, but sometimes reaching the height of a small tree. It is among the most common and widely distributed Cape proteas, preferring lower altitude sites with drier northern and northwestern exposures. Its flowers are about three inches (8 cm) in diameter and typically have bright red bracts. Like king protea, sugarbush has many local varieties, some of which are in bloom at any time of the year.

Ray-Flowered Protea. Ray-flowered protea (*Protea eximia*) is a common protea that also grows in large, dense stands in mountain terrain of the southern Cape and reaches the height of a small tree. Flower heads have widely spaced red bracts that fade to cream at the base. Its broad leaves have a fine red margin (Fig. 7.08C).

Long-Leaved Protea. Long-leaved protea (*Protea longifolia*) is found close to sea level in the mild areas of the southwestern Cape. Its flower heads are most commonly pink but are sometimes white (Fig. 7.08B), red, or purple-black. Bot River protea (*P. compacta*),

A

B

C

D

E

F

G

H

FIG. 7.08. *Examples of the genus* Protea. *A. King protea* (Protea cynaroides) *is the floral symbol of South Africa. B. Long-leaved protea* (P. longifolia) *is named for its elongated leaves. The flower heads in this variety have pale green bracts around a feathery center. C. Ray-flowered protea* (P. eximia) *has widely spaced bracts that fade to cream at the base. The leaves have a fine, red margin. D. Mountain rose* (P. nana) *has relatively small, red, bell-shaped flowers on downward turning stems. E. Sunshine protea* (Leucadendron xanthoconus) *gives a yellow color to coastal slopes of fynbos in early spring. F. Red bottlebrush* (Mimetes cucullatus) *has flower heads on tall, erect branches that stand out over the surrounding fynbos vegetation. Orange pin-cushion protea* (Leucospermum cordifolium) *(G.) and L. reflexum (H.) have brightly colored styles.*

which is usually pink, grows near the town of Hermanus. Mountain rose (*P. nana*) grows in dense stands up to an elevation of 3,000 feet (900m) and has relatively small, red, pendulous, bell-shaped flowers (Fig. 7.08D).

Sunshine Proteas. *Leucadendron* is also a genus of the protea family that includes shrubs and trees. An example of the latter is the silver tree (*L. argenteum*), which grows to 40 feet (12 meters) and develops a distinctively fissured gray trunk with age. Its leaves are covered by silky, silvery hairs that reflect light. In dry or windy weather, its hairs lie flat, minimizing water loss from stomata. In damp weather, hairs stand out like fur, trapping water droplets from moist air. This attractive tree is restricted almost entirely to the Cape Peninsula, and large numbers are found at the Kirstenbosch National Botanical Garden. *L. xanthoconus* is a smaller member of this genus, growing to three to six feet (1 to 2 m) (Fig. 7.08E). Several species of *Leucadendron* color many fynbos mountain slopes yellow in the winter and early spring. It is the bracts surrounding the flower heads on male plants that turn the brightest color. Female plants have flowers with woody bracts that give them a cone-like appearance.

The Genus Mimetes. The most common and widespread species of this small genus of the protea family is red bottlebrush (*Mimetes cucullatus*). Red bottlebrush has tall and striking, bright red flower heads at the top of long, erect branches (Fig. 7.08F).

Pincushion Proteas. Pincushion proteas in the genus *Leucospermum* are named for the rings of brightly colored yellow, orange, or red styles that protrude like curved pins from their flower heads. The tallest species in the genus grow to heights of a small tree. Orange pincushion protea (*L. cordifolium*) (Fig. 7.08G) is pollinated by Cape sugarbirds that are attracted to its abundant nectar, particularly during the peak bloom period of September to October. *L. reflexum* is another species in this genus (Fig. 7.08H).

THE GENUS ERICA

The heath family (Ericaceae) has a virtually worldwide distribution, but its largest genera are concentrated in small areas. For example, of about 1,200 species of *Rhododendron*, more than 700 are in the area where China, Burma, and India meet. Similarly, of over 740 species in the genus *Erica*, 650 are endemic to South Africa (Schuman and Kirsten, 1992).

Heaths of the genus *Erica* are found only in Africa, Europe, and a small part of western Asia. Their distribution has a long north-south axis from northern Norway to the southernmost portion of the Cape, where the remarkably large number of 625 species are found. These are mainly near the coast and on the coast-facing, southern and southeastern slopes of the mountains where rainfall is highest. The highest concentration of *Erica* species and maybe the highest species density of a single genus in the world is in the district around the town of Calendon, about 70 miles (120 km) east of Cape Town, where there are over 235 species (Schumann and Kirsten, 1992).

Ericas are thought to have originated in Africa after the breakup of Gondwana (page 26), since they are not found in Australia or South America. The distribution into Africa and Europe could have arisen by bridging the Mediterranean Sea and the relatively temperate chain of East African highlands. Surprisingly, one species of *Erica* is even found in the once moister mountains of the Sahara Desert and the Arabian peninsula. Europe is home to about 20 species, most of which grow around the Mediterranean Basin.

Ericas range from ground-hugging shrubs to a few species that grow as tall as 20 feet (6 m). They generally thrive in acid soil and some species can tolerate standing water in bogs and marshes. A common feature of most species is their short, narrow, needle-like leaves whose edges are turned downwards, leaving only a narrow slit on the lower surface (Fig. 7.09). The stomata or pores lie within the slit on the underside of the leaves, reducing loss of moisture. The upper surface of the leaves is hard and waxy.

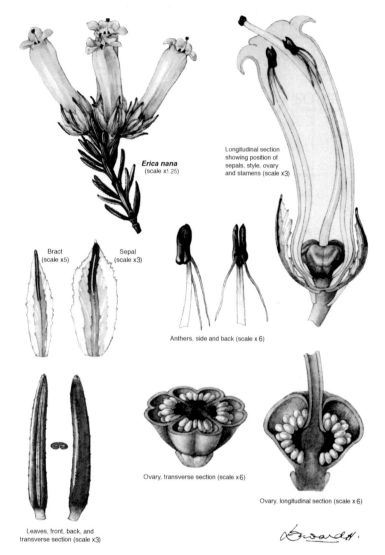

Erica nana
(scale x1.25)

Bract
(scale x5)

Sepal
(scale x3)

Longitudinal section showing position of sepals, style, ovary and stamens (scale x3)

Anthers, side and back (scale x 6)

Ovary, transverse section (scale x6)

Ovary, longitudinal section (scale x 6)

Leaves, front, back, and transverse section (scale x3)

FIG. 7.09. *Erica structure. Leaves are short, narrow, and needle-like, with edges that are turned downwards, leaving only a narrow slit on the lower surface. The stomata or pores through which these plants exchange gases and lose water lie within the slit on the underside of the leaves, reducing loss of moisture during periods of drought. The upper surface of the leaves is hard and waxy. Flowers are bell-shaped or tubular, with four joined petals and eight stamens. (From a drawing by E. Ward-Hilhorst in Schumann et al., 1992, with permission from Fernwood Press.)*

Flowers are bell-shaped or tubular, typically with four joined petals and eight stamens. Colors range from white through yellow, orange, and pink to red (Fig. 7.10).

About 80 percent of *Erica* species are insect pollinated (Schumann et al., 1992), but not primarily by bees, which are not particularly common in fynbos (Johnson, 1992). Many

A

B

C

FIG. 7.10. *Three species of* Erica: A. E. massonii, B. E. patersonia, *and* C. E. coccinea *var.* pubescens.

species of *Erica* are pollinated by flies that have a long proboscis. The flowers of these species typically have concentrated nectar within a flask-shaped flower with a narrow opening, flanked by a landing pad for the insect. The close fit between the structure of the flower and its pollinator is believed to have come about through coevolution, a process by which interacting plant and animal species undergo *reciprocal* changes. Flies with a long proboscis are at a selective advantage in this interaction, since this helps them to better reach the nectar from a tubular flower (Fig. 7.11). Longer tubular flowers have an analogous reproductive advantage by forcing a fly to press more deeply into a flower in search of nectar, resulting in a passive pick up or deposit of pollen. The reciprocal selection for longer flowers and flies with a longer proboscis has been called an example of runaway coevolution (Johnson, 1992). However, this degree of coevolution is considered to be rare. Whereas specific plants commonly appear to have adapted to pollinators, there are fewer examples of a *mutual* adaptation of pollinators to plants.

FIG. 7.11. *Coevolution. Close fits between the structure of flowers and the anatomy of pollinating animals can come about through coevolution, a process by which interacting plant and animal species undergo reciprocal changes. Flies with a long proboscis are at a selective advantage in this interaction, since this helps them to reach more nectar from a tubular flower. (Reproduced with permission of Colin Patterson-Jones, Cowling and Richardson, 1995.)*

About 15 percent of *Erica* species are bird pollinated, chiefly by the orange breasted sunbird, whose long, curved beak has an appropriate shape for long, tubular flowers. These species have thicker stems to support the weight of the bird. The remaining five percent of species are wind-pollinated and have small, inconspicuous, bell-shaped flowers that release clouds of pollen when shaken.

FIG. 7.12. *Three Ericas from copper plate engravings that appeared in* Curtis's Botanical Magazine *between 1792 and 1797. All three were named by Carl Linnaeus, the originator of the binomial system of nomenclature, by which plants are identified by genus and species. On the left is* E. baccans, *a common heather on the Cape Peninsula with pink, urn-shaped flowers. On the right is* E. grandiflora, *with large, tubular flowers whose color varies according to district from yellow to orange and red. In the center is the fire heath,* E. cerinthoides, *which has pink to red tubular flowers. It is the most common and best-known species of* Erica *in the Western Cape. Its common name comes from its ability to regrow after a fire from persistent root stock.*

Erica species are able to adapt to the often highly leached, nutrient-poor soils of the Western Cape by virtue of their mutually beneficial or symbiotic relationship with fungi. These extend from their roots far into the surrounding soil, helping to extract soil nutrients such as phosphorus and nitrogen more efficiently (page 39).

Seeds of many species of Western Cape *Erica* were taken to Europe during a period of intensive plant exploration at the end of the 18th and early in the 19th centuries. During this age of sailing ships with their scant supplies of fresh water, seeds could be more easily and successfully transported than living plants. The successful propagation of many species of *Erica* from seed is described and illustrated in early issues of *Curtis's Botanical Magazine*, published with hand-colored prints in England from 1787 until 1983.

Three illustrations of common species of *Erica* appeared in the late 1700s in *Curtis's Botanical Magazine*, and are reproduced in Fig. 7.12. All three were named by Carl Linnaeus, the originator of the binomial system of nomenclature by which each plant is identified by a genus and a species name. Linnaeus, a professor at Uppsala, Sweden, never visited South Africa; however, Thunberg, one of his students, spent three years between 1772 and 1775 collecting plants from the region around Cape Town. He sent thousands of specimens back to Sweden for study. *Erica cerinthoides*, commonly known as fire heath, is probably the best known and most widespread South African species of *Erica*. Its tubular flowers are pink to red, but its name comes not from the color but from its vigorous sprouting from a surviving root stock after a fire. Fire heath grows to shoulder height with leaves and clusters of flowers at the ends of its branches. *E. grandiflora* grows to a similar height and also has tubular flowers, ranging from orange to dark red, according to the region where they

A

B

FIG. 7.13 A. and B. Restionaceae. The rush-like Restionaceae are the most common plants of the fynbos, where they take the place of grasses and give a brown-green hue to the landscape. They grow in dense clumps and have tough stalks with wiry leaves.

grow. *E. baccans* grows to the height of a tall shrub with abundant, cup- or bell-shaped flowers. It colors whole slopes of the Cape Peninsula a deep pink during its spring bloom season.

Restios. Restios (Restionaceae) are a family of reed or rush-like plants that are prominent in the fynbos and strandveld, where they take the place of grasses and give a brown-green hue to the landscape (Fig. 7.13). The Restionaceae has 12 genera and 310 species, of which 290 are endemic to the fynbos region. The family is also common in Australia and is represented in Chile, thus having a Gondwana-type distribution. It is not found in the Mediterranean Basin or California. Restios grow in dense clumps and have tough stalks with wiry leaves. Taller species are about waist height, but there are also many lower-growing varieties. Each clump is unisexual, with male and female plants often having dissimilar appearances. Male plants of the same genus tend to be very similar, making their identification difficult (Heywood, 1978).

Other Plants. Other plants that are common in the fynbos include geophytes or bulbs, numerous members of the daisy family (*Asteraceae*), and pelargoniums (of the family Geraniaceae), which are the source of window-box "geraniums." These will be discussed next under the heading of Strandveld, where their variety and abundance are greatest.

Strandveld

Strandveld is a coastal plant community that occurs along a narrow strip of well-drained, sandy terrain along the Atlantic coast north of Cape Town (McMahon and Fraser, 1988) (Fig. 7.04). The vegetation of this coastal region is also described as fynbos/thicket mosaic, dune thicket, or succulent thicket (Cowling and Holmes, 1992), and it occurs adjacent to larger areas of fynbos. Annual rainfall for areas of strandveld vegetation is near the dry end of the range for mediterranean climate areas, between about five and 15 inches (12.5-35 cm) in the coastal plain. This is augmented by moisture from coastal fog.

Strandveld consists of hard-leaved evergreen shrubs, bulbs, many plants of the daisy family (Asteraceae), and fleshy-leaved ice plants commonly known in South Africa as vygies (in the large Mesembryanthemum complex). A period of spectacular bloom peaks in the spring months of September and October. The rest of the year, the landscape is dominated by the brown-green color of the abundant restios. By contrast to the enormous plant diversity of the fynbos, the strandveld supports a more moderate number of species with a smaller proportion of endemics. The number and diversity of bulbs and members of the daisy family is an outstanding feature.

Bulbs. The mediterranean climate area of the Western Cape is home to the greatest variety of geophytes or bulbs in the world (Bryan, 1989; Jeppe and Duncan, 1989; Doutt, 1994), and many of these are found in the strandveld. Bulbs account for an estimated quarter of the species in the mediterranean climate part of the Western Cape.

Iris Family. The iris family (Iridaceae) is particularly well represented, even though Cape Province has no members of the genus *Iris*. Among the more familiar members of this family, many of which have been hybridized and cultivated, are *Gladiolus, Ixia, Freesia, Sparaxis, Babiana,* and *Watsonia* (Fig. 7.14). Many were introduced into gardens in the mediterranean climate area of Western Australia, where climate and soil conditions are very close to those of the Western Cape. Interestingly, several of the South African species that have become rare in their home territory have become highly invasive in Australia,

Fɪɢ. 7.14. *Geophytes or bulbs. The amaryllis family is represented by A.* Haemanthus coccinea. *In the iris family are B.* Watsonia galpinii, *C.* Sparaxis elegans, *and D.* Babiana pulchra. *E. The* Kniphofia *sp. is in the lily family.*

A

B

FIG. 7.15. A. *Namaqualand daisies (Ursinia cakilefolia) in Skilpad Wildflower Reserve near Kamieskroon in semi-arid Namaqualand. B. Cape strawflower (Phaenocoma prolifera) in Fernkloof Nature Reserve near Hermanus on the south coast. C. Daisies in West Coast National Park.*

C

where they are now considered weeds. Bulbs bloom most abundantly after fires and in disturbed or cultivated soil.

South African species of *Gladiolus* are commonly known as afrikanders, pypies, or painted ladies. They have fewer flowers than the typical hybridized garden gladiolus. The name gladiolus is derived from the Latin gladius, meaning sword, after the sword-like leaves. Watsonias have a superficial resemblance to gladioli. Their pink, orange, red, or white blooms are at their peak in late spring.

Amaryllis Family. The amaryllis family (Amaryllidaceae) typically has large bulbs that are well adapted to store enough food and water to survive through a hot, dry summer. In summer or early fall, flowers in most plants of the amaryllis family are produced before leaves, giving them a somewhat naked appearance (Fig. 7.14A). Their leaves remain green through the rainy season and replenish food supplies to be stored in the bulb for the next season. The belladonna lily (*Amaryllis belladonna*) and many species of *Haemanthus* share this characteristic, a common one in the Western Cape but rare in other mediterranean climate regions. The summer-blooming *Agapanthus* spp., known as Christmas candle in South Africa, is also in the amaryllis family. This name is inappropriate in the northern hemisphere, where it is called lily of the Nile.

Among the bulbs that bloom prolifically in response to fire (Doutt, 1994) is the fire lily (*Cyrtanthus ventricosus*). It blooms in just a few weeks following a fire, even though it is outside its normal flowering season. *Haemanthus* is in another genus of the amaryllis family that blooms quickly after a fire. Summer and fall wildfires also stimulate early flowering of *Watsonia borbonica*, which normally blooms in the spring.

Lily Family. The lily family (Liliaceae) includes red hot poker, another strandveld bulb that has become a popular garden plant in mild winter regions (Fig. 7.14E). *Kniphofia uvaria* is probably the parent of most of these.

Calla lily (*Zantedeschia aethiopica*), in the large Araceae family, is known as arum lily in South Africa. This popular garden plant originated in the strandveld, where pigs were once allowed to feed on its rhizomes. It grows in marshy places, where it blooms during the winter rainy season.

Daisy Family

The daisy family (Asteraceae) is well represented in strandveld (Fig. 7.15A-B) and further north into semi-arid Namaqualand. There, daisies dominate the landscape to an unparalleled degree during August and September, if winter rains have been adequate. The most dramatic carpets of flowers are in abandoned wheatfields, such as those in the Skilpad Wildflower Reserve near the town of Kamieskroon. Fig. 7.15A shows a field colonized primarily by orange-flowering *Ursinia cakilefolia*. *Ursinia* does not require insects for pollination and insects are, in fact, scarce in such fields (Smuts and Bond, 1995). Adjacent unploughed land just across a narrow lane has a much more diverse flora with varied and abundant insects. It appears that ploughing of natural veld eliminates numerous plant species that do not readily reinvade even when the fields are left fallow for several years. Only annuals that do not require pollination by insects successfully invade adjacent abandoned farmlands. The contrast between ploughed and unploughed fields raises difficult questions regarding conservation. Species-poor fields of brightly colored flowers are clearly a major attraction for most tourists; however, less dramatic and more undisturbed veld is of far greater botanical interest and preserves an important plant habitat.

Woodland and Forest

Trees and forests have not been a prominent part of the mediterranean climate landscape of the Western Cape in historic times (Specht and Moll, 1983). They were once found mainly on steep hillsides and near watercourses, and some remain in these locations today (Figs. 4.05D and 4.06C). Van Riebeek, the Dutch founder of the Cape Colony, described areas of "forests with thousands of fairly tall and straight trees" from which "one could get thousands of complete masts for ships" (McMahon and Fraser, 1988). Within 50 years of this first Dutch settlement at Cape Town in 1652, most of the lower slopes of Table Mountain had been cleared of good-sized trees. Wood was needed not only for fuel and construction by the settlers; it was vital for ship repairs. The virtual deforestation of the area around Cape Town resulted from provisioning a highly profitable trade in Asian spices, porcelain, and silk that was a basis for Dutch prosperity in the 17th and 18th centuries.

After the Napoleonic Wars and before the opening of the Suez Canal in 1889, Cape Town under British rule remained the major provisioning point for ships on the long voyage between Europe and the Far East. During this period, several species of *Eucalyptus* were brought from Australia. More recently Monterey pine (*Pinus radiata*), maritime pine (*P. pinaster*), and Aleppo pine (*P. halepensis*) have become common plantation trees. These, as well as Australian species of *Acacia* and *Hakea*, are invasive in areas of native fynbos (Richardson et al., 1992).

The invasiveness of foreign trees may be surprising, since indigenous trees are so rare in the Cape (Richardson et al., 1992). One explanation is that many of the invasive species are very effective in dispersing their stored canopy seeds after fires, a characteristic not as common among the indigenous trees. Most of the native trees are referred to as having afromontane origins, because they are closely related to species in the more tropical East African highlands. Many of these trees have fleshy fruits that impose high demands for soil nutrients. Furthermore, they lack the mycorrhizal association with root fungi that benefits an estimated three-quarters of fynbos species by a more efficient extraction of nutrients from poor soils (Cowling, 1992).

Near Cape Town, there are still areas of native forest between Kirstenbosch Botanical Garden and the top of Table Mountain that include groves of silver tree (*Leucadendron argenteum*). The silver tree has become a popular ornamental tree, prized for its silvery leaves, which shimmer in the wind. The Cape chestnut (*Calodendrum capense*) is also a tree of the Cape Peninsula that is found in a few remaining groves and that has become a popular ornamental. A common tree that grows along streams is wild almond (*Brabejum stellatifolium*), which is in the protea family. Its poisonous fruit becomes edible after soaking and was used as a coffee substitute by early settlers. A hedge of wild almond was planted in 1660 to keep in livestock and mark the eastern edge of the Cape settlement. Part of this hedge, now consisting of broad, dense trees, is still preserved in the Kirstenbosch Botanical Garden.

The largest remaining forests in the Western Cape occur behind the coastal towns of Knysna and George, about 240 miles (400 km) east of Cape Town. These temperate rain forests are on the south-facing slopes of the Outeniqua and Tsitsikamma Mountains, in an area of year-round rain that is slightly east of the mediterranean climate region of the province but still within the eastern limits of the area of fynbos. The forests include the yellow-wood and stinkwood and were the home of the last remaining elephants in the region.

Yellowwood (*Podocarpus latifolius*) is among the tallest trees, growing to 160 feet (50 meters). It is a slow-growing conifer, which is prized for its easily worked wood that can be polished to an attractive yellow sheen. Stinkwood (*Ocotea bullata*), despite its unappealing name, has strong, dark wood, which was used to make most of the furniture during the early period of settlement. It is widely used as a street tree, grows to 20 feet (6 meters), and has large clusters of pale pink, five-petalled flowers. Ironwood (*Olea capensis*) was one of the few trees that remained uncut because it was so difficult to fell.

There are also sheltered slopes on the Hottentots Holland Mountains and Langeberg where the proteas *P. nitida* and *P. laurifolia* and *Maytenus oleoides* reach a tree-like 20 feet (6 m) if protected from fire. North of Cape Town, in the Cedarberg Mountains, are scattered groves of the Clanwilliam cedar (*Widdringtonia cedarbergensis*) from which the range derived its name.

PLANTS FOUND IN MEDITERRANEAN CLIMATE AREAS OF WESTERN CAPE PROVINCE, SOUTH AFRICA

Agapanthus spp.	lily of the Nile	Amaryllidaceae	bulbous
Asteraceae spp.	daisy family	Asteraceae	herb/shrub
Babiana spp.	baboon flower	Iridaceae	bulbous
Brabejum stellatifolium	wild almond	Proteaceae	evergreen shrub/tree
Erica spp.	heath	Ericaceae	evergreen shrub
Gladiolus spp.	afrikander	Iridaceae	bulbous
Homeria spp.	Cape tulip	Iridaceae	bulbous
Ixia spp.	ixia	Iridaceae	bulbous
Kniphofia spp.	red-hot poker	Liliaceae	bulbous
Leucadendron argenteum	silver tree	Proteaceae	evergreen shrub/tree
Leucospermum spp.	pincushion	Proteaceae	evergreen shrub
Mesembryanthemum	vygie	Aizoaceae	succulent
Olea capensis	ironwood	Oleaceae	tree
Ocotea bullata	stinkwood	Lauraceae	tree
Pelargonium spp.	geranium	Geraneaceae	evergreen shrub
Protea cynaroides	king protea	Proteaceae	evergreen shrub
Protea eximia	ray-flowered protea	Proteaceae	evergreen shrub
Protea neriifolia	oleander-leaved protea	Proteaceae	evergreen shrub
Protea repens	sugarbush	Proteaceae	evergreen shrub
Restionaceae spp.	restio	Restionaceae	rush-like herb
Watsonia spp.	pypie	Iridaceae	bulbous
Zantedeschia aethiopica	calla or arum lily	Araceae	bulbous

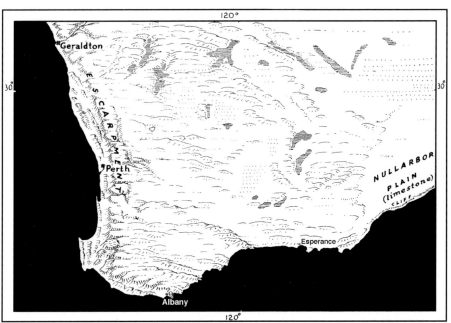

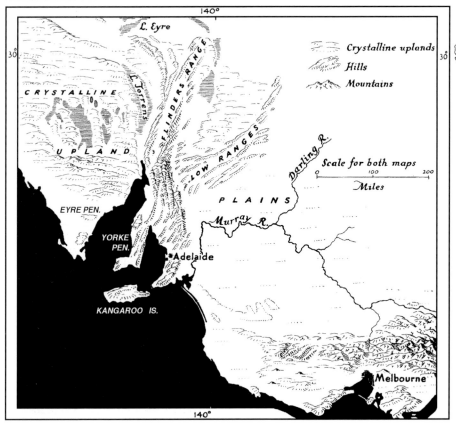

8.
AUSTRALIA

LANDSCAPE AND CLIMATE

T here are two mediterranean climate areas in Australia, the southwestern part of the state of Western Australia and the southernmost part of the state of South Australia. These areas are small in relation to the continent as a whole and are separated by the broad expanse of the semi-arid to arid Nullarbor Plain (Fig. 8.01). Western Australia has the greater diversity of plant species. The mediterranean climate part of South Australia has small extensions into adjacent parts of Victoria and New South Wales. Plants of South Australia's mediterranean flora are shared to a large extent with adjoining areas in the eastern states, even at the species level. Elevation is a less important influence on climate and vegetation than in any of the other mediterranean climate regions, since there are no high mountains, and even mountains of modest size are few and far between.

Western Australia

Western Australia is Australia's largest state, occupying an area that is about the size of the continental United States west of the Rocky Mountains. Most of the state consists of desert and is thinly populated. Almost all of the population of about 1.5 million is concentrated in the moister southwestern tenth of the state that has a mediterranean climate. Perth, the capital and only large city, accounts for two-thirds of the entire state's population with just over a million people. It is the most geographically isolated city of its size in the world.

The mediterranean climate area of Western Australia corresponds closely to the distinctive and varied plant community of the Southwest Botanical Province. This province covers a triangular area of about 100,000 square miles (300,000 km²), extending from the northern extreme of Shark Bay on the Indian Ocean to about 350 miles (550 km) east of Albany on the southern coast (Beard, 1982) (Fig. 8.02).

FIG. 8.01. *Topographical maps including the mediterranean climate regions of Western Australia above and South Australia below. Orientation maps are on the right. Abbreviations are for the Australian states: WA for Western Australia; SA, South Australia; Q, Queensland; NSW, New South Wales; and V, Victoria. NT stands for Northern Territories. (Adapted from a map by Erwin Raisz with permission of John Wiley & Sons.)*

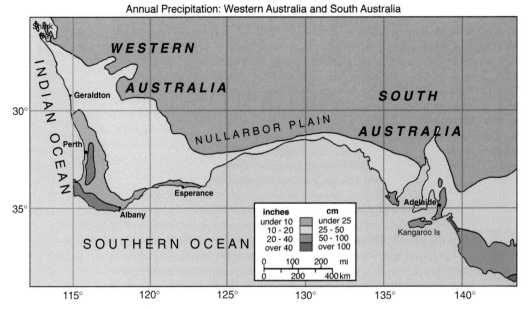

FIG. 8.02. *Annual rainfall. In Western Australia, rainfall is heaviest toward the southwestern extreme of the state. (Pate and Beard, 1984.) In South Australia, rainfall is greatest in the south. (Based on Griffin and McCaskill, 1986.)*

South Australia

South Australia is slightly larger than the combined areas of California, Nevada, and Arizona. The driest parts of the entire continent occur in the north. But the southernmost sixth of the state, an area of about 60,000 square miles (180,000 square km), has a relatively mild, mediterranean climate similar to that of southwestern Australia. As in Western Australia, this is where most of the population of about 1.5 million is concentrated, primarily in the state capital Adelaide, which like Perth has a population of about one million. South Australia's mediterranean climate region extends from Adelaide south and east to the border with the state of Victoria, Kangaroo Island to the southwest, and the Yorke and Eyre peninsulas to the west. North of Adelaide, drier portions of the mediterranean climate region extend into the Flinders Ranges which rise to about 3,900 feet (1,200 m). Although the adjacent northwestern extreme of Victoria and the southwestern corner of New South Wales also have a mediterranean climate, the larger parts of these states, including the capitals of Melbourne and Sydney, have year-round rainfall.

Rainfall

The southwestern tip of Western Australia is the most stormy and rainy area of the state. Rainfall decreases further inland and toward Perth and Geraldton to the north (Fig. 8.02). Since winter temperatures are mild and there are no high mountains, snow is a rarity. As a result, streams are highest in the winter, in contrast to California and Chile where spring and early summer snowmelt in the high mountains accounts for the greatest stream flow. Rainfall typically comes in heavy storms that are concentrated in the middle

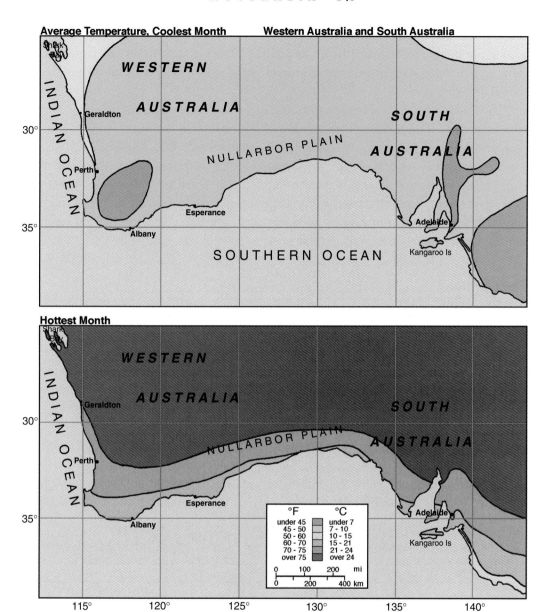

Average Temperature, Coolest Month   Western Australia and South Australia

Hottest Month

°F	°C
under 45	under 7
45 - 50	7 - 10
50 - 60	10 - 15
60 - 70	15 - 21
70 - 75	21 - 24
over 75	over 24

FIG. 8.03. *Average temperature of the hottest and coolest months, January and July, respectively.* *(Based on Watt, 1940.)*

of the winter, as in California, Chile, and the Western Cape. These come as westerly gales, and together with strong sea breezes in the summer, account for the fact that Perth is the windiest state capital in Australia. From Perth north to Geraldton and beyond, summer rainfall is rare, but the area around Albany in the south has significant summer rains. The highest mountains of the Stirling Range north of Albany (3,640 feet or 1,110 m) and the ranges of hills like Darling Scarp just east of Perth are moister than surrounding plains.

No part of South Australia has as much rainfall as the extreme southwestern tip of Western Australia, thus South Australia has almost no tall forests analogous to those of

Western Australia. Summer thunderstorms are not infrequent in South Australia, though most rain is still concentrated in the winter months. However, winter rain storms are typically less intense than in southwestern Australia, and are more likely to be spread out over several days (Fox, 1995).

The mediterranean climate parts of Western Australia and South Australia, as well as most other parts of the continent, are notable for the substantial variability in rainfall and the high frequency of drought. Consequently, large wildfires are common (AUSLIG, 1990). The frequency is about once per 20 years in the moistest parts of South Australia, with most fires occurring in summer and autumn. Interestingly, there are fires at more frequent intervals of three to five years along the more populated coastal southeast of the continent between the two largest cities, Sydney and Melbourne, where they are more likely to result from human activities than from lightning.

Temperature

The mediterranean climate area of Western Australia extends slightly closer to the equator than any of the other mediterranean climate regions. The mediterranean climate areas of both states have relatively mild and uniform minimum temperatures in winter (Fig. 8.03) with hot summers. Perth has heat waves accompanied by dry winds from the northeast desert interior of the state, corresponding to the Santa Ana winds in Southern California.

PLANT COMMUNITIES

Kwongan is a Western Australian form of sclerophyll scrub vegetation that corresponds closely to the fynbos of the Western Cape in South Africa. Kwongan is found mainly along

FIG. 8.04. *Mediterranean plant communities in Australia. (Based on Specht, 1981; Sparrow, 1989; AUSLIG, 1990; Beard, 1990.)*

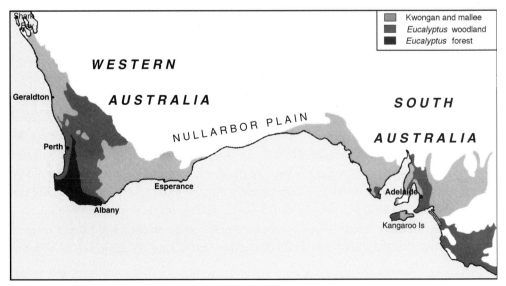

the Indian Ocean between Perth and Shark Bay and near the Southern Ocean between Albany and Esperance (Fig. 8.04). It extends inland for scarcely more than 60 miles (100 km) with little in the way of mountain barriers. In South Australia, there are similar but less extensive forms of scrub vegetation with scattered low trees, thickets, and heath, but the term scrub heath is used for these plant communities. These kwongan-like forms of vegetation are found in the national parks and nature reserves on Kangaroo Island, on the Mount Lofty Ranges near Adelaide, and in the southeastern part of the state, adjacent to Victoria.

Another common form of scrub vegetation is mallee, which is named for the multi-trunk forms of *Eucalyptus* that can sprout from underground lignotubers after a fire. Mallee is most extensive in the drier, northern half of the mediterranean climate region of South Australia that lies north of Adelaide (Sparrow, 1989). It is also found in Western Australia, inland from and mixed with kwongan.

Woodlands dominated by *Eucalyptus* are most extensive in the parts of Western Australia where rainfall is greater than where kwongan and mallee grow. The tallest and densest forests of *Eucalyptus* are in the highest rainfall areas south of Perth and near the south coast, toward Albany. Woodlands and forests are far less extensive in South Australia, but occur on the western half of Kangaroo Island and in a few protected areas of the Adelaide Hills.

In both of Australia's mediterranean climate regions, the greatest loss of native vegetation in recent decades has resulted from the replacement of woodland and mallee by wheat farming. The area devoted to wheat farming almost trebled in the 1960s and 70s (AUSLIG, 1990). The expansion was made possible by the recognition that agricultural productivity could be dramatically enhanced by the addition of fertilizer containing trace minerals, copper and zinc, that were deficient in the natural soil. More land has also been cleared for grazing livestock because new fertilizers made it possible to raise rich forage grasslands in areas that were not previously considered suitable for agriculture. These changes have been most striking where annual rainfall is between 10 and 20 inches (25 and 50 cm). In landscapes of pastures and crops, native vegetation often remains only along roads, near scattered clumps of trees, and around granite outcrops.

Dominant Families and Genera

Before further discussion of specific plant communities, it seems appropriate to introduce the three major plant families that dominate not only major forms of vegetation in the mediterranean climate areas, but the Australian continent as a whole. These are the myrtle (Myrtaceae), pea (Fabaceae), and protea families (Proteaceae). Just the two large genera *Eucalyptus* and *Acacia* give the landscape a distinctively Australian appearance. These two genera account for the canopy plants in three-quarters of Australia (AUSLIG, 1990).

GENERA WITH AN ABUNDANCE OF SPECIES

Vegetation in southwestern Australia is particularly noted for its richness of species, a total of about 8,000 within a relatively small number of families and genera (Hopper, 1992). *Acacia* alone accounts for more than 400 species and *Eucalyptus* for over 300. Just in the protea family, there are more than 200 species of *Grevillea*, 100 *Hakea*, nearly 50 *Dryandra*, and 50 *Banksia*. In the case of *Banksia*, for example, there are only 14 species in the rest of the continent (George, 1986).

The pattern of abundant species but few genera is just the opposite from that found in the rainforest of Queensland, on the northeastern edge of the continent, where there is a great diversity of families and genera that have a paucity of species. The difference provides a clue as to changes in southwestern Australia over the last 40 million years. The fossil record indicates that the southwest, along with most of the rest of the continent, had a subtropical climate with a broad-leaved rainforest resembling that found on the coast of Queensland today (White, 1986). During the subsequent 30 million years or so, the wet, closed forest was gradually replaced by open forest, woodlands, and heath, as cycles of aridity became increasingly frequent and severe (Barlow, 1994).

The extinction of many families and genera and the explosion of speciation among the survivors are thought to have taken place during drastic climate changes over the last 40 million years. These culminated in a series of ice ages and polar glacial melts over the last two million years (Dodson, 1994). It is during this period that southwestern Australia developed a mediterranean climate and was cut off from the rest of Australia by arid climates to the north and east. During these ice ages, the arid center of the continent expanded, numerous species became extinct, and sclerophyll forests were pushed toward the coast. With the return of a warmer climate, forests and shrubs recovered but with a different mix of species each time. A pattern resembling the present vegetation did not begin until about 13,000 years ago, or 5,000 years after the last glacial maximum. In the southwest, for example, an increase in moisture between 9,000 and 7,000 years ago favored karri (*Eucalyptus diversicolor*) forest at the expense of jarrah forest (*E. marginata*).

Myrtle Family

The myrtle family is characterized by trees and shrubs with aromatic leaves.

Eucalypts. Eucalypts grow best in the moister parts of the continent, including the two mediterranean climate regions, where they dominate the canopy of virtually all woodlands and forests (Figs. 4.05B, 4.06E). They are well adapted to nutrient-poor soils by virtue of mycorrhizal associations that have been found in all species examined (Pryor, 1976).

The mature, strap-like leaves of most eucalyptus trees hang vertically, which avoids

A

the drying effects of direct sunlight and allows considerable sunlight to penetrate to the ground. This enables a rich variety of understory plants to grow. The flowers of eucalypts have a hard lid or operculum that drops off when the flowers open, revealing multiple stamens as the most colorful and prominent feature. The mottlecah tree (*E. macrocarpa*), a mallee growth form native to Western

B

Australia, is notable for having the largest flowers in the genus, with a diameter of four inches (10 cm). The fruit is a fire-resistant, woody capsule.

Fig. 8.05. *Karri forest (Eucalyptus diversicolor). Even though the trees are closely spaced, eucalypts let enough light through the canopy to support a rich variety of understory plants. These include the colorful vines (A.) tree hovea (Hovea elliptica) and (B.) coral vine (Kennedia coccinea).*

A

B

Bottlebrushes. Many genera of the myrtle family in addition to *Eucalyptus* are native to Australia. Among these are four commonly known as bottlebrushes that are found in the mediterranean climate region of Western Australia. The common name, bottlebrush comes from the brushlike appearance of the cylindrical or globular flower clusters in which the stamens are most prominent. One-sided bottlebrushes *Calothamnus* and *Eremaea* are both endemic to Western Australia. More than half of the 24 species of *Kun-*

Fɪɢ. 8.06. A. *Paperbark* (Melaleuca *sp.*) *growing in waterlogged soil in early spring near Perth. B. Yellow flower of bell-fruited mallee* (Eucalyptus preissiana), *which is found between Albany and Esperance in Western Australia. C. Yellow puffball-like flowers of gold dust wattle* (Acacia acinacea) *endemic to Kangaroo Island, South Australia (opposite).*

zea are also endemic to Western Australia. The fourth bottlebrush genus *Callistemon* is mostly from eastern Australia, but is also found in Western and South Australia.

Tea Trees and Paperbarks. Tea trees and paperbarks represent two important genera in the myrtle family. Tea trees (*Leptospermum*) and paperbarks (*Melaleuca*) range from low shrubs to trees (Fig. 8.06A). The genus *Leptospermum* includes the silver tea tree, *L. sericum*, which has the largest flowers of the genus (pale pink and 1-2 inches or 2 to 5 cm across) and is found along the southern coast of Western Australia in the area of Esperance.

Shrubs of the Myrtle Family. Shrubs of the myrtle family include two genera, *Verticordia* and *Darwinia*, most of whose species are endemic to the mediterranean climate region of southwestern Australia. *Verticordia* are commonly known as featherflowers for their feathery or woolly clusters of colorful flowers and are common in the kwongan of the northern sandplains near Geraldton. *Darwinia* is one of the notable genera of flowers of the Stirling Range north of Albany, where eight species are endemic. The shrubs have a heathery appearance with small, colorful, bell-shaped flowers. Different species are found on each peak of this range, presumably as a result of speciation when the peaks in the Stirling Range were islands.

Pea Family

The pea family as a whole is well represented worldwide. Most of the species in this family are well adapted to the nutrient-poor soils of Australia by virtue of their association with nitrogen-fixing bacteria in their root nodules. These bacteria can convert nitrogen from the air to nitrogenous compounds that can be used by the host plant.

Wattles. Wattles (*Acacia*) belong to the pea sub-family (Mimosaceae) and are among the world's largest genera with about 1,200 species. *Acacia* is common in the mediterranean climate parts of Australia, and it dominates the vegetation in semi-arid areas. Members of the genus *Acacia* are referred to as wattles in Australia and typically bear yellow, puffy flowers (Fig. 8.06C) with numerous stamen. Seeds are borne in pods similar to peas and beans. In contrast to African *Acacia*, most of which have compound leaves and thorns, Australian species often have phyllodes or expanded leaf stalks instead of true leaves and rarely have thorns.

C

Vines. The pea sub-family Papilionaceae contains numerous vines that are prominent in western Australia. They are characterized by their asymmetrical flowers, with an upper wing-shaped petal and a broad lower keel petal. Seeds develop in a pod. Western Australia alone has 52 genera in this sub-family. Some are prominent as bright-flowered understory vines in tall karri (*Eucalyptus diversicolor*) forests and include early-blooming, purple-flowered bushes of *Hovea*, masses of tangled, climbing vines of native wisteria (*Hardenbergia comptoniana*), and low-growing, red coral vine, *Kennedia coccinea* (Fig. 8.05B). Several species in the genus *Gastrolobium* have yellow and brown flowers and are notorious for their toxicity to livestock resulting from the poison monofluoroacetate. These species were responsible for catastrophic losses of farm animals in the early days of colonization, which accounts for common names such as York Road poison (*G. calycinum*) and crinkle-leaf poison (*G. villosum*). Forty-five of the 47 species of *Gastrolobium* are found in Western Australia.

Protea Family

The protea family (Proteaceae) is as prominent in Australia as it is in the Western Cape, South Africa. The greatest diversity of genera and species in this family occurs in the southwestern corner of the continent.

Banksias. *Banksia* is Australia's best-known genus of Proteaceae, with its large cylindrical or cone-like, compound flowers (Fig. 8.07).

Other Large Genera. Other large genera (Fig. 8.11) include *Dryandra*, which is endemic to Western Australia and also has large compound flowers with prickly leaves. *Grevillea* typically has red, yellow, orange, or white tubular flowers. Cone bushes (*Isopogon*) are usually shrubs with prickly leaves, often with yellow or pink flowers. Smoke bush (*Conospermum*) is named for its masses of woolly flowers found on the periphery of shrubs that form hemispheres roughly three feet (1 m) in diameter.

Kwongan and Scrub-Heath

Kwongan is a word in the aboriginal Nyungar language of southwestern Australia that refers to open, scrubby vegetation on sandy soil (Fig. 4.02D). A similar though less diverse form of vegetation in South Australia is known as scrub-heath. Kwongan resembles shrublands of other medi-

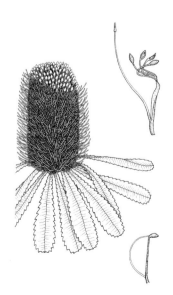

FIG. 8.07. Banksia *flower structure. The* Banksia *are Australia's best-known genus of Proteaceae. The large cylindrical head of B.* serrata *bears compound flowers, each of which has a stigma and an anther that splits into four cup-shaped, pollen-bearing perianth tips. (Reproduced with permission, Wrigley and Fagg, 1989.)*

A

terranean climate areas including chaparral, matorral, and maquis (Pate and Beard, 1984; Beard, 1990;
Fox, 1995). It has most in common with the fynbos of
South Africa in respect to its open spacing of shrubs,
abundance of understory plants, and richness in plant
species. Kwongan is at its most spectacular in the spring
months of September to November, when shrubs and
understory herbs are in their peak bloom. But the period of growth and flowering extends well into the dry
summer, as it does in the fynbos of the Western Cape.

*The Protea Family (Banksia, Grevillea, Dryan-
dra, and Hakea) and the Eucalypts.* In the driest, northernmost areas of its distribution, kwongan consists mainly of shrubs whose height varies with the time elapsed
since the last fire. South of Perth, where rainfall is greater,
kwongan includes a denser growth of tall shrubs and low
trees including the genus *Banksia* (Fig. 8.08A). Some of
these reach a height of about 20 feet (6 m). The shrubs

B

Fɪɢ. 8.08. *Banksia is in the protea family, as is the African
genus* Protea, *which has similarly large flower heads. A. and B.
Scarlet banksia (B.* coccinea) *in the Stirling Range near Albany.
C. Silver banksia (Banksia* marginata) *from South Australia.*

C

FIRE

Australian shrubs are designated as *sprouters* or *seeders*, according to their response to a fire, as in other mediterranean areas of chaparral-like vegetation. Most prominent among the sprouters are *Eucalyptus* called mallee (page 159). Other sprouters with lignotubers are the sheoak, *Casuarina pusilla* and the tea tree, *Leptospermum myrsinoides*. Both of these have underground root crowns that can sprout when a fire has destroyed the above-ground portion of the plant. Another form of sprouting is exemplified by the grass tree, *Xanthorrhoea australis*, which has a caudex or reduced underground stem that is buried about 15 inches (35 cm) under the soil surface. The vegetative apex of the caudex sprouts after a fire.

Plants that are known as seeders are in the minority in Australian heathlands. When the vegetation was studied 10 years after a fire in the heathland of the moistest part of South Australia, 95 percent of the plants by weight

FIG. 8.09. Banksia *cones: seed capsules are open on the top cone and closed on the bottom one.*

were sprouters and only five percent had regenerated from seed (Specht, 1994). Seeder plants have no root crowns or burls that bud and are easily killed by fire. One mode of survival for them is a woody seed capsule that remains on the plant for many years (Fig. 8.09). Capsules are split open by a fire and release seeds, which germinate on fire-denuded but nutrient-enriched soil. Among such seeders are *Banksia ornata*, *B. marginata*, *Hakea rostrata*, *H. rugosa*, and *Casuarina muellerana*. In another method of seed dispersal, seeds retain hard, woody coats for many years as they lie dormant on the ground until a fire cracks their seed coats.

A third mode of seed dispersal, also common in South Africa (page 123), employs harvester ants, which are attracted to plants with seeds bearing a lipid-rich appendage called an elaiosome. Ants carry these seeds away from the parent plant to their underground nests. Among Australian heathland plants, roughly 1,500 species in 87 genera and 24 families rely on this reproductive strategy (Specht, 1994).

In the case of many species of *Eucalyptus*, harvester ants will eat all seeds as they gradually drop to the ground (Parsons, 1994). After a fire, however, more seeds are dropped than the ants can eat. Much of this excess is stored in underground ant nests where uneaten seeds germinate with winter rains.

of the genera *Grevillea*, *Dryandra*, and *Hakea* and a highly varied group of perennial and annual herbaceous plants are also prominent. The openness of the kwongan structure probably makes fires less frequent here than in other forms of scrub and woodland vegetation.

Along the southern coast, kwongan is found in the Stirling Range and the coastal hills called the Barrens or the Barren Range. The highest peak of the Barren Range is only 1,475 feet (450 m), but it nevertheless has a dramatic appearance, situated next to the coast. Here, shrubs and herbs are combined with gnarled, taller mallee species (*Eucalyptus tetragona*, *E. marginata*, or *E. staeri*). The Stirling Range has an unusually high number of endemic species, including the scarlet banksia (*Banksia coccinea*) (Fig. 8.08A, B). Some of the vegetation of this range has a close physical resemblance to the mountain fynbos of the Western Cape though it is botanically different.

Everlastings. Although perennials account for most of the spring wildflowers of Western Australia, the annual everlastings (including *Helichrysum* and *Helipterum*) form carpets of white, yellow, or pink flowers that are especially luxuriant after a wet winter (Figs. 3.16 and 8.14B). These genera of the daisy family have over 100 species, most of which are represented in either southwestern Australia or the Western Cape.

UNUSUAL POLLINATORS

A large number of flowers in Western Australia are bird-pollinated (Morcombe, 1968). Many are in the protea family and include species of *Banksia*, *Dryandra*, *Grevillea*, and *Hakea* with stems that are strong enough to support the weight of a bird. The honeyeater, for example, with its long, curved beak, is ideally suited both to gather nectar and to receive pollen from the tubular *Grevillea* flower (Fig. 8.10A).

A

The role of marsupials in pollination is probably unique to Australia (Hopper et al., 1996). The small, mouse-size, nocturnal honey possum (*Tarsipes rostratus*) is well suited to feeding on nectar and digesting pollen from certain members of the Protea family (Fig. 8.10B). The close relationship between plants and their animal pollinators makes the plants vulnerable, because populations of some of these animals is decreasing through competition with newly introduced species, such as rabbits and feral cats.

B

Fig. 8.10. *Unusual pollinators. A. A honeyeater receiving pollen from the style of a Grevillea flower as it drinks nectar from the base of the flower. B. A small nocturnal honey possum, a mouse-sized marsupial, picks up pollen from the protruding styles of a Banksia flower as it drinks the abundant nectar. (Reproduced with permission of Fagg, Wrigley and Fagg, 1989.)*

VARIATIONS IN SPECIES OVER SHORT DISTANCES

I n Western Australia as in the Western Cape, there are areas where numerous species have evolved within short distances. In a region of scrub vegetation, such as the Mount Lesueur area near Eneabba, north of Perth, one can walk a mere 0.3 mile (0.5 km) and find that more than 60 percent of the plant species will be different (Hopper, 1992; Hopper et al., 1996). A similar degree of diversity can be found along the Darling Escarpment east of Perth and among the individual peaks of the Stirling Range near Albany. In these places, species typically occur as small populations lacking in good means of seed dispersal. The species diversity exists in sites where it cannot be attributed primarily to climatic or topographic barriers. Many species in these areas are characterized by a high degree of natural hybridization, an indication of increased evolutionary activity. It is therefore not surprising that the mediterranean climate region of Western Australia has the highest number of rare and endangered species in Australia (Hopper, 1992). The large number of species in certain genera makes plant identification difficult except by specialists.

While vegetation of the mediterranean climate part of South Australia does not have the enormous diversity of species of the southwest, dominant genera are the same (Specht, 1972). There are 34 species of Eucalyptus and 32 species of Acacia in the mediterranean part of South Australia. These figures are only about one-tenth of the number for the same genera in southwestern Australia (Hopper, 1992).

Though there is a desert barrier of the Nullarbor Plain between the two states, 40 percent of South Australia's species are nevertheless shared with Western Australia. The part of South Australia with a mediterranean climate also has many species in common with areas in the eastern states that do not have rain concentrated in the winter. A surprisingly high 80 percent is shared with the eastern states of Victoria, New South Wales, and Queensland (Specht, 1972). Probably this is because there is no major climate or topographic barrier in that direction (Dashorst and Jessop, 1990).

In South Australia there are fewer endemic or endangered plants than in Western Australia. However, three areas in the mediterranean climate part of the state each have 40 or more rare or endangered plants: the Adelaide Plains and Hills, the Eyre Peninsula to the west, and Kangaroo Island to the southwest (Briggs and Leigh, 1988). By comparison, the number of rare or endangered plants in Western Australia ranges between 177 and 337 for each of six geographic subdivisions of the mediterranean climate part of the state.

Sundews. Abundance of the sundew family (Droseraceae) is a characteristic feature of kwongan and of southwestern Australia generally. The leaves of this insectivorous family of annual and perennial herbs have hairs tipped with clear, sticky, insect-trapping droplets

A

B

C

D

FIG. 8.11. *Members of the protea family from four genera: A.* Dryandra nivea *near Perth (photo by Lester Hawkins, reproduced by permission of Maggie Wych, Western Hill Nursery). B. Prickly plume grevillea (*Grevillea annulifera*) near Geraldton. C. Common smoke bush (*Conospermum stoechadis*) north of Geraldton. D. Sundews, such as* Drosera whittakeri, *trap insects on their sticky leaves. E. Cone bush (*Isopogon ceratophyllus*) from Kangaroo Island, South Australia.*

(Fig. 8.11D). Plants are adapted to nutrient-poor soils by digesting trapped insects as a source of nutrients. Many grow as rosettes of leaves on the ground, often around granite outcrops; other can climb to a height of six feet (2 m).

Large granite outcrops are common in the southwestern part of Western Australia. Some form large sheets of rock that emerge only slightly above the surrounding land, whereas others form dramatic mounds, as in the Porongurup Range near Albany (Fig. 8.14A). Often, largely bare rocks have patches of everlastings, lichens, and mosses. Soil buildup in cracks and depressions create gardens of plants that tolerate alternate waterlogging and desiccation. The edges of the outcrops have increased moisture from runoff during the rainy season, often supporting the growth of *Drosera*, *Stylidium*, and orchids.

E

A

B

FIG. 8.12. *Grass trees (Xanthorrhoea sp.). A. Closeup of a flower stalk with many minute white flowers. B. Landscape with grass trees, one of which has a tall, narrow, brown flower stalk.*

Grass Tree. The grass tree (*Xanthorrhoea* sp.) (Fig. 8.12) has a woody trunk topped with a dense head of grass-like leaves. The trunk is often black from earlier fires, which stimulate the emergence of a tall, flowering spike with numerous, minute white flowers. Nurseries take advantage of this fire-dependent growth and flowering by offering to burn off old leaves with a blowtorch when the plant is sold.

Restios. These rush-like perennials of the Restionaceae family are prominent in the shrublands of Western Australia, but are not as dominant as in the fynbos of the Western Cape. As in the Cape, male and female plants grow in separate clumps.

Australian Heath and the Citrus Family. Scrub vegetation in South Australia includes members of the Australian heath family, Epacridaceae, which resembles the true heaths or Ericaceae that are most varied in the Western Cape. Typically, the Epacridaceae are shrubs with small leaves and tubular flowers, many of which bloom in the winter (Fig. 8.13A, B). The citrus family (Rutaceae) is also well represented by two Australian genera, *Correa* and *Boronia* (Fig. 8.13C, D).

Fanflowers and Kangaroo Paws. The right-of-way along roadsides is relatively wide in Australia and is among the best places to see wildflowers when driving through farm-

A

B

land. Figs. 8.15B and D show two colorful roadside examples of the fan-flower family (Goodeniaceae) in the genus *Lechenaultia*. The family has 13 genera with about 300 species, of which more than two-thirds are found in Western Australia. The kangaroo paw family (Haemodoraceae) includes Mangle's kangaroo paw (*Anigozanthos manglesii*), which has become the floral emblem of Western Australia. It has a striking color combination of a hairy, felt-like red stem and brilliant red and green flowers (Fig. 8.15C). The genus *Anigozanthos* is endemic to Western Australia. Other of its species can be recognized from their similar claw-shaped, hairy flowers and unusual color combinations. The genus *Conostylis* is in the same family (Fig. 8.15A).

C

Mallee

Mallee refers to a growth form of the genus *Eucalyptus* and to the unique Australian plant community that it dominates. About one-fifth of the species of *Eucalyptus* sprout to form multiple trunks from thick, horizontal, underground rootstocks or lignotubers. The trunks grow to the height of a tall shrub or small tree (Parsons, 1994). Mallee-forming species of *Eucalyptus* are referred to as bull mallee or big mallee if there are only a few large trunks and as

Fig. 8.13. *Roadside shrubs in spring bloom on Kangaroo Island, South Australia. A. Common heath* (Epacris impressa) *and B. flame heath* (Astroloma conostephoides) *are both in the Australian heath family, Epacridaceae, which resembles the true heaths in the family Ericaceae. C. Salmon correa* (Correa pulchella) *and D. island boronia* (Boronia edwardsii) *are both in Australian genera of the citrus family, Rutaceae.*

D

A

FIG. 8.14. A. *Large granite outcrops of the Porongurup Range with the Stirling Range in the distance, both a one-hour drive north of Albany. B. Strawflowers or everlastings of the daisy family in bloom in early spring nearly covering a granite outcrop west of Albany.*

B

whip-stick mallee if there are many thin ones. Stems branch at the top and bear leaves only at the ends of branches, giving the appearance of a low canopy. Fires are frequent in mallee, but the large water- and nutrient-storing lignotubers survive and soon sprout new stems.

The plant community of mallee, like other forms of scrub vegetation in mediterranean climate habitats, is found in drier areas than woodlands and forests. On the most nutrient-poor soils, mallee can resemble kwongan, with a diverse understory of heath-like vegetation referred to as mallee heath. These are the areas in which there is a striking bloom of wildflowers in the spring. On richer soils, the understory consists primarily of grasses and herbaceous plants including the spinach family, Chenopodiaceae (Specht, 1981; Sparrow, 1989).

B

C

A

Some of the same mallee species of *Eucalyptus* are prominent in both South Australia and Western Australia, including *Eucalyptus incrassata*, *E. oleosa*, and *E. socialis*. The mallee-forming cup gum (*Eucalyptus cosmophylla*) is endemic to South Australia and is found only on Kangaroo Island and the Mount Lofty Range near Adelaide (Fig. 8.16). Similarly concentrated on Kangaroo Island is a tall form of mallee, narrow leaf mallee (*Eucalyptus cneorifolia*) (Fig. 8.17A).

In parts of South Australia with nutrient-poor soils and less rain, *Eucalyptus incrassata* becomes common. In the driest mallee areas, the prickly hummocks of porcupine grass (*Triodia irritans*) also become prominent. These hummocks grow outward to a diameter of three feet (1 m), eventually leaving a dead center. Leaves come to sharp points, which readily penetrate the skin. Another unpleasant characteristic of the plant is marked flammability, which results in extremely fast-moving and therefore highly dangerous wildfires.

D

FIG. 8.15. *Roadside plants in Western Australia. A. A species of* Conostylis, *which belongs to the same family,* Haemodoraceae, *as the genus* Anigozanthos. *B.* Lechenaultia formosa *north of Perth (photo by the late Lester Hawkins, reproduced with permission of Maggie Wych, Western Hills Nursery). C. Mangle's kangaroo paw (Anigozanthos manglesii) is the floral emblem of Western Australia, to which the entire genus is endemic. D. Blue lechenaultia (L. biloba) near the town of New Norcia, north of Perth.*

Woodland and Forest—Western Australia

The largest forests in southwestern Australia are dominated by either of two endemic species of *Eucalyptus*, commonly known as jarrah (*E. marginata*) and karri (*E. diversicolor*).

Jarrah Forest

The tallest jarrah trees grow near streams and are as high as 100 feet (30 meters). They grow in almost pure stands (Fig. 8.17B) or mixed with one or two other eucalypt species,

---

INVASIVE PLANTS

Introduced members of the iris family have been notoriously invasive in Western Australia, where native bulbs are rare compared with other mediterranean climate areas. For example, *Gladiolus caryophyllaceus*, an endangered species in its native South Africa, has escaped from gardens in southwestern Australia to become an aggressive weed. Other invasive plants include species of *Watsonia*, *Babiana*, *Freesia*, *Homeria*, *Sparaxis, and Ixia*, many of which garderners have to coax to survive in other parts of the world. Roadside gladioli and freesias have become commonplace in Western Australia and are often the targets of herbicide treatment.

Climate and soil conditions are sufficiently similar in mediterranean climate regions of Australia and the Western Cape to help explain the degree of invasiveness of some of the plants that have been exchanged between the two continents. Invasive species are freed from native plant competitors and insects that keep them in check on their home territory. Introduced species are abundant near drainage channels and roadsides where nutrients accumulate and in agricultural areas, where soils are enriched by fertilizers.

In the state of South Australia, about 20 percent of a flora of more than 3,000 species is exotic (Specht, 1972). Major contributions to the exotic flora come from the Mediterranean Basin, Europe, Western Asia, South Africa, Chile, and California. Invasive plants are among the most successful weeds in newly disturbed soil and in plowed land. One of the most notorious exotic plants in South Australia is Paterson's curse (*Echium plantagineum*), a garden plant introduced from the Mediterranean Basin in the 1850s. It spread rapidly enough to be recognized as a problem in the 1880s and is now widespread in the eastern states, especially New South Wales. It is an annual or biennial herb with intense blue or violet flowers and is poisonous to livestock if eaten in large quantities. Another highly invasive plant is the South African daisy (*Senecio pterophorus*), which now dominates much of the understory of woodlands in the Adelaide Hills. As in Western Australia, showy South African bulbs imported as garden plants have escaped to outnumber native bulbs, particularly along roadsides and in wooded areas.

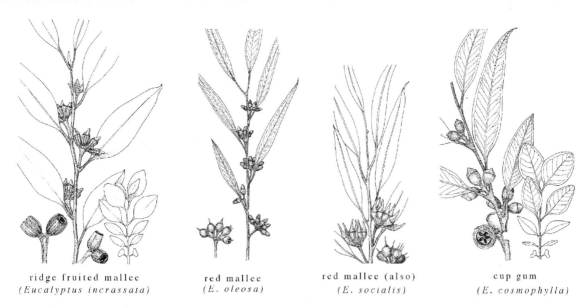

ridge fruited mallee   red mallee   red mallee (also)   cup gum
(*Eucalyptus incrassata*)   (*E. oleosa*)   (*E. socialis*)   (*E. cosmophylla*)

FIG. 8.16. *Eucalyptus trees can be most easily identified by the shape of their fruit. Four common species that can grow as mallee forms are shown in drawings reproduced by courtesy of the Botanic Gardens of Adelaide, South Australia. Species found in South Australia and in Western Australia include E. incrassata, E. oleosa, and E. socialis. The cup gum (E. cosmophylla) is endemic to South Australia. Note that the juvenile leaves of E. incrassata and E. cosmophylla are different in shape from those of the mature tree.*

marri (*E. calophylla*) and/or wandoo (*E. wandoo*). Jarrah wood has an attractive red-brown color and darkens with age. It is durable, fire-resistant, strong, and easily worked by wood-crafters. Understandably, it has been highly prized and was once the primary source of paneling in railway sleeper cars throughout Australia. Now, though supplies are more limited, it is still used for expensive flooring, furniture, and wood turning. Most of the old growth jarrah forest has been logged, and what remains consists mainly of younger trees.

Understory Plants. Understory plants such as *Banksia*, *Casuarina*, and *Melaleuca* thrive in *Eucalyptus* forests because the vertically positioned leaves allow light to penetrate.

Sheoak. Sheoak is in the genus *Casuarina*, which has 30 species, 29 of which are found only in Australia. The trees have vestigial leaves which are on thin, segmented, green stems with a superficial resemblance to pine needles (Fig. 8.18). The common name, sheoak, comes from an oak-like appearance of the wood. There are separate male and female plants, the female being more easily distinguishable by its hard, woody seeds. The male has a single brown stamen in its minute flower.

Paperbark. Paperbark (*Melaleuca* spp.) (Fig. 8.18) is in the myrtle family and has white, yellow, or pink flowers, which are conspicuous in the summer and fall when relatively few other plants, except for those in the myrtle family, are in bloom (Fox, 1995). Paperbarks are adapted to grow in low areas where soil becomes waterlogged following winter rains (Fig. 8.06A) but is baked dry in the summer.

About two decades ago, a serious disease appeared afflicting patches of some *Eucalyptus* forests and their understory plants (Specht, 1981). In Western Australia, jarrah forests

A

were among the most vulnerable, with understory plants of the Proteaceae family also extremely susceptible to what proved to be a root fungus disease (Wrigley and Fagg, 1991). The responsible organism, *Phytophthora cinnamomi*, spreads easily, and effective methods of control have not been discovered.

Karri Forest

Karri forest (*Eucalyptus diversicolor*) is the tallest and most magnificent forest of Western Australia (Figs. 4.06E, 8.17C). It is found only in a small, moist area of the southwesternmost tip of the state where annual rainfall is greater than 40 inches (100 cm). Some trees

FIG. 8.17. A. *The narrow leaf mallee* (Eucalyptus cneorifolia) *on the eastern part of Kangaroo Island is among the tallest forms of mallee. B. Jarrah* (E. marginata) *forest with tall, brown trunks south of Perth. Most jarrah forest is now second growth. C. Karri* (E. diversicolor) *(opposite) forms the tallest forest in Western Australia. The straight trunks are nearly white.*

B

Fig. 8.18. Two common genera of Australian trees. The she-oak, Casuarina stricta, is found in South Australia and in states to the east. The paperbark, Melaleuca lanceolata, is widely distributed in Australia. (Reproduced courtesy of the Botanic Gardens of Adelaide, South Australia.)

drooping sheoak
(*Casuarina stricta*)

dry land tea tree
(*Melaleuca lanceolata*)

grow to a height of 250 feet (75 m) and a trunk diameter of six feet (2 m). The species name, *diversicolor*, refers to the color difference between the deep green upper surface and the paler green lower surface of the leaves. Tree trunks are smooth and nearly white.

Understory Plants. Trees in the understory include karri oak (*Casuarina decussata*), peppermint (*Agonis flexuosa*), and bull banksia (*Banksia grandis*). A shrub layer may include the karri wattle (*Acacia pentadenia*). After a severe fire, acacia and other members of the pea family are among the first to appear along with karri seedlings, which can grow to as high as 80 feet (25 m) in 20 years. Colorful vines are common in the understory of karri and jarrah forests. They include the low-growing coral vine (*Kennedia coccinea*) and tall masses of the blue tree hovea (*Hovea elliptica*) (Fig. 8.05A, B). Purple *Hardenbergia* may climb low shrubs and trees in tangled masses. White *Clematis pubescens* is sometimes conspicuous climbing on tree trunks.

There is an enormous variety of orchids in Western Australia with 23 genera and numerous species, many of which grow in the forest of the southwestern part of the state (Fig. 8.19 A-C).

Woodland and Forest—South Australia

In South Australia, forest habitat is scant. Scattered woodlands are dominated by species

C

A

B

C

FIG. 8.19. *Forest orchids. A. Spider orchid (Caladenia patersonii). B. Cow-slip orchid (Caladenia flava) with lichen. C. Donkey orchid (Diuris sp.).*

of *Eucalyptus*, but few of them are endemic to the state. The river red gum (*Eucalyptus camaldulensis*) (Fig. 8.20), for example, is found throughout most of Australia. It is a picturesque broad, spreading tree that is found mainly along streams and is described as Australia's best-loved tree. In the summer, it has abundant clusters of white flowers. Messmate stringy bark (*E. obliqua*) (Fig. 4.05B) and brown stringy bark (*E. baxteri*) are also found in the Mount Lofty Range and on Kangaroo Island. The range of both species extends into the state of Victoria, to the east. Eucalyptus trees can often be identified by the fruit on the lower branches or on the ground.

FIG. 8.20. *South Australian Eucalyptus trees. (Reproduced courtesy of the Botanic Gardens of Adelaide, South Australia.)*

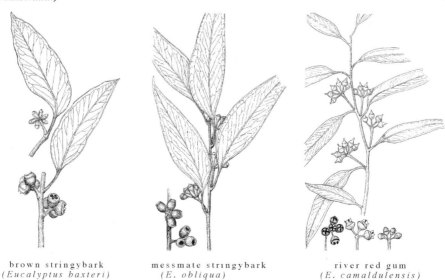

brown stringybark
(*Eucalyptus baxteri*)

messmate stringybark
(*E. obliqua*)

river red gum
(*E. camaldulensis*)

Plants Found in Mediterranean Climate Areas of Australia

Acacia spp.	wattle	Fabaceae	evergreen tree/shrub
Anigozanthus manglesii	Mangle's kangaroo paw	Haemodoraceae	perennial herb
Banksia coccinea	scarlet banksia	Proteaceae	evergreen shrub
Banksia marginata	silver banksia	Proteaceae	evergreen tree/shrub
Caladenia patersonii	spider orchid	Orchidaceae	bulbous
Callistemon speciosus	Albany bottlebrush	Myrtaceae	evergreen shrub
Casuarina fraseriana	common sheoak	Casuarinaceae	evergreen tree/shrub
Conospermum stoechadis	common smokebush	Proteaceae	evergreen shrub
Correa pulchella	salmon correa	Rutaceae	evergreen shrub
Darwinia spp.	darwinia	Myrtaceae	evergreen shrub
Diuris spp.	donkey orchid	Orchidaceae	bulbous
Drosera spp.	sundew	Droseraceae	herb
Epacris impressa	common heath	Epacridiaceae	evergreen shrub
Eucalyptus baxteri	stringy bark gum	Myrtaceae	evergreen tree
Eucalyptus calmaldulensis	river gum	Myrtaceae	evergreen tree
Eucalyptus diversicolor	karri	Myrtaceae	evergreen tree
Eucalyptus marginata	jarrah	Myrtaceae	evergreen tree
Grevillea spp.	grevillea	Proteaceae	evergreen shrub
Hakea spp.	hakea	Proteaceae	shrub
Hardenbergia spp.	native lilac	Fabaceae	vine
Helichrysum spp.	everlastings	Asteraceae	annual
Helipterum spp.	everlastings	Asteraceae	annual
Isopogon ceratophyllus	cone bush	Proteaceae	evergreen shrub
Kennedia coccinea	coral vine	Fabaceae	vine
Lechenaultia biloba	blue lechenaultia	Goodeniaceae	perennial shrub
Melaleuca spp.	paperbark	Myrtaceae	shrub/tree
Nuytsia floribunda	W.A. Christmas tree	Loranthaceae	semi-parasitic tree
Xanthorrhoea preissii	grass tree	Xanthorrhoeaceae	shrub/tree

9.01. Topography of the Mediterranean Basin. Coastal areas are mainly mountainous with few coastal plains, most of which are narrow. (Adapted from a map by Erwin Raisz. Reproduced with permission of Rand McNally.)

9.

THE MEDITERRANEAN BASIN

LANDSCAPE AND CLIMATE

T*he Mediterranean Basin has a far larger area and more complex geography than any of* the other mediterranean climate regions (Fig. 9.01). In the north, its land mass is divided by the Mediterranean Sea into three large peninsulas: the Iberian peninsula to the west; the Italian boot in the center; and the Balkan peninsula to the east. Asian Turkey lies to the northeast. Lining its eastern shore are Syria, Lebanon, and Israel. The southern coast of the Mediterranean Sea has a less convoluted shape, and the countries of the Maghreb (Morocco, Algeria, and Tunisia) are cut off from the Middle East by a broad band of Libyan and Egyptian desert. The mediterranean climate portion of this region is primarily coastal, with larger inland extensions in Spain and Turkey.

Rainfall

The pattern of rainfall in the Mediterranean Basin is extremely varied, reflecting its large area and complex geography (Fig. 9.02). In general, rainfall is greater in the north than in the south, and it increases with elevation in the numerous mountain ranges throughout the region. Moist air blows east from the Atlantic Ocean in the winter half of the year. As a result, western parts of the Iberian, Italian, and Balkan peninsulas have more rain than corresponding eastern parts. The wettest parts of the Mediterranean countries include northern Portugal, northwestern Spain, northern Italy, western Greece, and southwestern Turkey. The driest areas include southeastern Spain and parts of northern Morocco and Algeria, and Israel. In North Africa, Egypt and most parts of Libya are too dry to support mediterranean vegetation.

The most unusual feature of the rainfall pattern in the Mediterranean Basin is that its seasonal timing varies more from one region to another than in any other mediterranean climate area (Fig. 9.03). Autumn is the rainiest season on the eastern coast of Spain, southern France, and northern and central Italy. If you plan an October trip to Rome on the basis of the reliably sunny California or Chile weather in the early autumn, for example, you are likely to be in for a rainy surprise. October is the wettest month of the year in central and northern Italy, bringing periodically destructive floods. In other areas, peak rainfall is in winter and spring, or autumn and winter, or spring, or spring and autumn. The

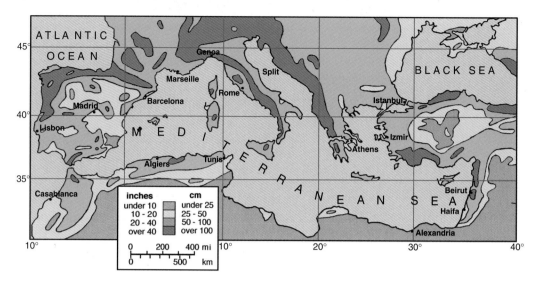

FIG. 9.02. *Annual precipitation in the Mediterranean Basin. In general, rainfall is greatest in the mountains and toward the north. The Iberian peninsula, Italy, and Greece are wetter in the west than in the east. (Based on Dudal, 1966.)*

more familiar mediterranean climate pattern of maximum rainfall in mid-winter prevails in coastal North Africa, southern Spain, Sicily, southern Greece, eastern Turkey, and along the eastern Mediterranean.

Throughout the Mediterranean Basin, summer is always the driest season. But summers in the western Mediterranean, including the Iberian Peninsula, southern France, and Italy, are not rainless. Late spring and/or early fall are often wet. Consequently, plants do

FIG. 9.03. *Variations in seasons of peak rainfall in the Mediterranean Basin. Although summers are dry throughout most of the Mediterranean Basin, peak rainfall in the other seasons differs substantially from one region to another. This variability is more marked than in any of the other mediterranean climate regions. (Based on Huttary, 1950.)*

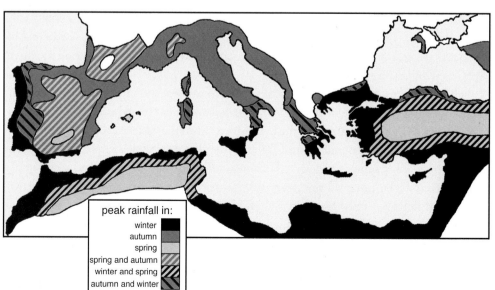

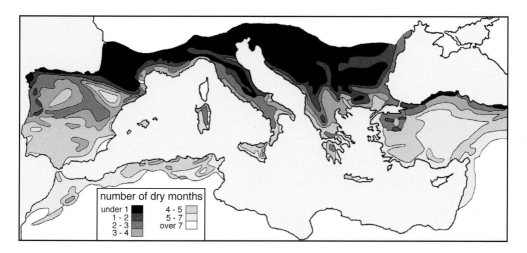

FIG. 9.04. *Duration of the dry season in the Mediterranean Basin. Duration of summer drought increases from north to south. The shortest dry season of less than one month is shown as black. Increasingly lighter grays and white depict the areas where the number of dry months are more numerous. (Based on Birot and Dresch, 1953.)*

FIG. 9.05. *Average temperature of the coolest month and hottest month, January and July, respectively. Winters are mildest along the coast. Coastal summer temperatures are quite high around the Mediterranean Sea because it has less of a cooling effect than the Pacific Ocean does in California and Chile (Fig. 5.03). (Based on Dudal et al., 1966; WMO, UNESCO, 1970.)*

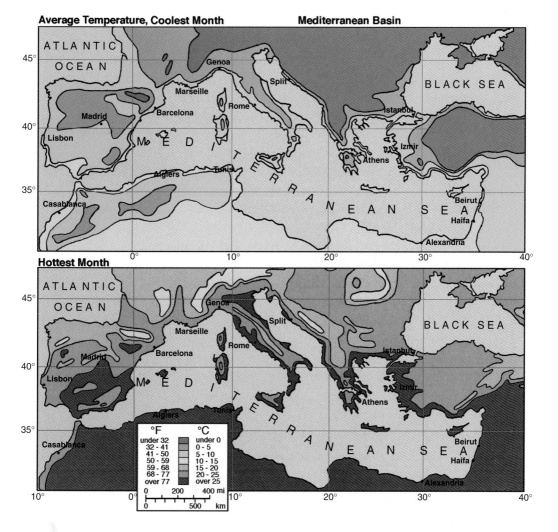

not have to endure six or more almost completely dry months, as in much of California and central Chile.

The average duration of summer drought varies considerably according to region in the Mediterranean Basin (Fig. 9.04) (Birot, 1964). Whereas most areas in the north have fewer than three dry months a year, much of the south has a dry period of five or more months. The eastern Mediterranean region, including Greece, Turkey, Lebanon, and Israel, also has a prolonged summer drought, which together with hot summers and mild

DISTINCTIVE FLORA AT THE MEETING OF CONTINENTS

The land masses of the Mediterranean Basin represent the adjoining portions of three much larger continents, Europe on the northern side of the Mediterranean Sea, Africa to the south, and Asia to the east and northeast (Fig. 9.01). Where the continents meet and where the Mediterranean Sea is narrowest, many species of plants are shared by two continents.

At the western extreme of the Mediterranean Sea, Spain is separated from Morocco by only eight miles (13 km) at the Strait of Gibraltar, which will be crossable by bridge or tunnel, according to current plans. The vegetation of the southern part of Spain and adjacent Portugal is given its special character by an estimated 480 species that are shared with North Africa (Polunin and Smythies, 1991). These include several species of *Narcissus*. Other common species with a natural distribution in the western African and European parts of the Mediterranean Basin are the cork oak (*Quercus suber*) and maritime pine (*Pinus pinaster*) (Fig. 9.06).

The Iberian peninsula, made up of Spain and Portugal, also has contact with France and Central Europe beyond. These geographical connections help to make it one of the regions of the Mediterranean Basin with the most diverse flora.

A second place where the mediterranean climate parts of North Africa and Europe are close is at the Sicilian Channel, which separates the eastern and western parts of the Mediterranean Sea. Here, between Tunisia and Sicily, the longest distance from land is only about 60 miles (100 km). In Europe, the dwarf fan palm (*Chamaerops humilis*) is found in the parts of Portugal, Spain, and Italy that lie closest to Morocco, Tunisia, and Algeria, where it has its major distribution (Fig. 9.06) (Merlo et al., 1993). It is found further north than any other native palm.

The richest flora of the Mediterranean Basin is found in Greece and the Balkans (Polunin, 1997). The proximity of this area to Asia accounts for more than 300 plant species of Asian origin. The Bosporus at Istanbul and the nearby Dardanelles separate the European and Asian parts of Turkey by only about one mile (1.6 km). The separation between the continents is now further diminished by two bridges carrying heavy traffic. Asian Turkey includes one of the largest areas of mediterranean climate and not surprisingly shares a great number of its species with Iran, Afghanistan, and Central Asia to the east. Examples of species restricted to parts of the eastern Mediterra-

winters produces a climate that is similar to the warmer and drier parts of California's mediterranean climate region.

Temperature

Summers in most of the Mediterranean Basin are hotter than they are in other mediterranean climate regions (Fig. 9.05) in part because the Mediterranean Sea becomes

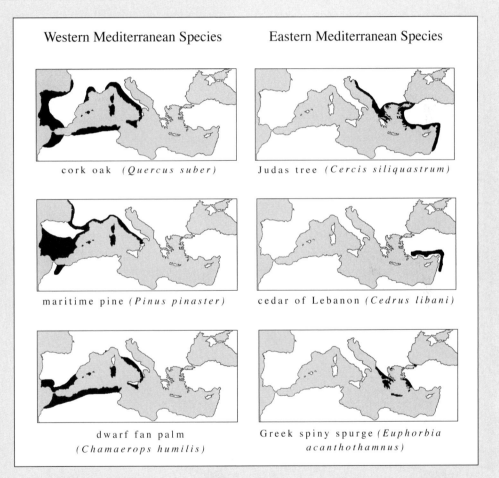

Western Mediterranean Species	Eastern Mediterranean Species
cork oak (*Quercus suber*)	Judas tree (*Cercis siliquastrum*)
maritime pine (*Pinus pinaster*)	cedar of Lebanon (*Cedrus libani*)
dwarf fan palm (*Chamaerops humilis*)	Greek spiny spurge (*Euphorbia acanthothamnus*)

FIG. 9.06. *Mediterranean plant species that are either eastern or western in their distribution. (Based on Daget, 1980.)*

nean include Judas tree (*Cercis siliquastrum*), cedar of Lebanon (*Cedrus libani*), and Greek spiny spurge (*Euphorbia acanthothamnus*) (Fig. 9.06).

An additional point of continental contact is represented by Israel, which lies near the junction of Africa and Asia, and accordingly shares species with both continents, as well as with Greece and other countries of Mediterranean Europe.

DISTRIBUTION OF PLANT SPECIES

Surprisingly many plant species are widely distributed in the Mediterranean Basin despite many water, mountain, and desert barriers (Fig. 9.07). Holm oak (*Quercus ilex*) is commonly used to define the extent of the region with a mediterranean climate and mediterranean-type vegetation. Various shrubs such as rock rose (*Cistus* spp.) cover similarly large areas. Other common species with extensive distributions are tree heather (*Erica arborea*), strawberry tree (*Arbutus unedo*), and Spanish lavender (*Lavandula stoechas*).

The range of the olive (*Olea europaea*) is used to define the extent of the mediterranean climate, like that of the holm oak. Its cultivation is believed to have been started along the eastern Mediterranean Basin by the ancient Phoenicians. It later spread to the west by Phoenician and Greek colonization. In Greek legend, the olive tree was given as a gift from the goddess Athena to the city of Athens (Baumann, 1993). Subsequently, the olive has played a central role in both the scenery and diet of the Mediterranean cultures. Olive trees take 25 years to come into full production, but they have a very long lifespan, gradually becoming more gnarled (Fig. 9.09). They sometimes exceed 1,000 years, but rarely reach 2,000 years in age (Mabberly and Placito, 1993).

FIG. 9.07. *Mediterranean plant species with extensive distributions. (Based on Daget, 1980.)*

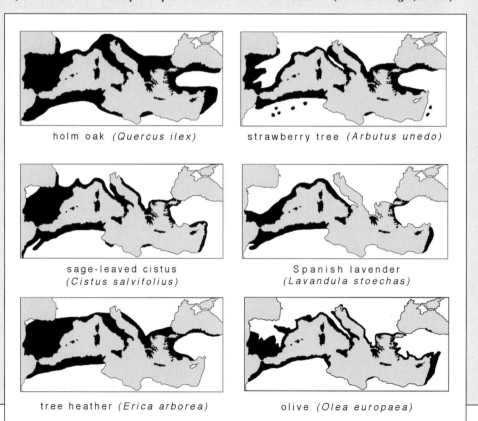

holm oak *(Quercus ilex)*

strawberry tree *(Arbutus unedo)*

sage-leaved cistus
(Cistus salvifolius)

Spanish lavender
(Lavandula stoechas)

tree heather *(Erica arborea)*

olive *(Olea europaea)*

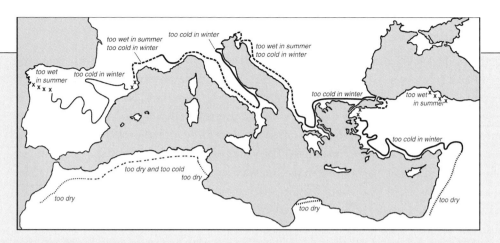

FIG. **9.08.** *The boundaries of the parts of the Mediterranean Basin in which olives are cultivated correspond closely to the extent of the mediterranean climate region. The climate conditions that limit the distribution of the olive are shown. (Adapted from Birot, 1953.)*

Boundaries for the cultivated olive are shown on Fig. 9.08, together with the climate conditions that limit its distribution (too dry, too wet in summer, too cold, etc.) (Birot, 1953). The olive requires 12 to 40 inches (30 to 100 cm) of annual rainfall and a humidity that is usually below 60 percent. The temperature can only rarely exceed 104°F (40°C) in the summer and cannot fall below 19°F (-7°C) for long in winter. The distribution of the olive is similar to that of the holm oak.

Another widely distributed tree is Aleppo pine (*Pinus halepensis*), which is particularly prominent in the hotter and drier eastern Mediterranean. Kermes oak (*Quercus coccifera*) is the common scrub oak of the Mediterranean Basin, most commonly growing as an evergreen shrub or small tree. Polunin and Huxley (1990) remark that where any two of the four (holm oak, olive, and Aleppo pine and Kermes oak) are seen growing together, "one can be pretty certain that one is in a Mediterranean climate."

FIG. **9.09.** *Olive trees (Olea europaea) that are over a century old take on picturesque, gnarled shapes.*

quite warm in the summer and provides less of a moderating effect on the climate than the oceans off California, Chile, and South Africa. The Mediterranean Sea is actually a giant basin whose inflow of fresh water from rivers such as the Rhone and Nile is exceeded by the evaporation of water from its surface. As a result, at the outlet of the Mediterranean Sea into the Atlantic Ocean through the narrow Strait of Gilbraltar, there is less water flowing out than cooler ocean water flowing in. The water temperature in the Mediterranean Sea rises from west to east, accounting in part for warmer temperatures in eastern coastal cities such as Athens, Izmir, and Haifa than in western cities such as Tunis and Barcelona. The temperature-moderating effect of the Atlantic Ocean is greatest near the western coastlines of Morocco and Portugal.

The mediterranean climate extends further toward the pole in the Mediterranean Basin than in any of the other five areas, to 45° latitude on the Italian Riviera. The mildness of the climate this far north results from the protection that the high Alps provide against cold northern weather. The Pyrenees Mountains between Spain and France create a similar barrier. Elsewhere, the Mediterranean Basin is less protected by mountain barriers. The effect of cool mistral wind from northern France and bora wind blowing south from Cen-

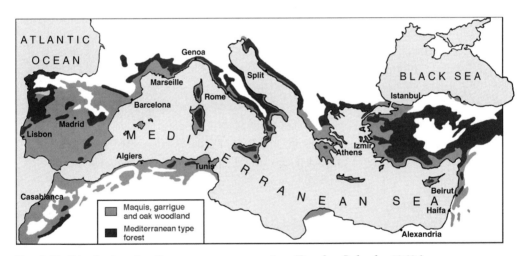

FIG. 9.10. *Distribution of mediterranean type vegetation. (Based on Lalande, 1968.)*

tral Europe limits the northward extension of some mediterranean plants. Hot summer winds, such as the sirocco that blows over southern Europe from the African Sahara desert, the etesian winds of Greece, and the meltemi that strike Turkey from central Russia, create similar conditions that limit less drought-tolerant plants.

PLANT COMMUNITIES

Maquis is a form of sclerophyll scrub vegetation that corresponds most closely to chaparral of California and matorral of Chile. It typically consists of tall, dense, evergreen shrubs and low trees. Maquis grows in areas with rainfall and temperatures that are intermediate between garrigue and sclerophyll woodland.

Garrigue is a sparser, lower form of vegetation with many aromatic shrubs that have soft foliage. Garrigue shares this characteristic with coastal scrub in other mediterranean climate regions. It generally occupies the drier, hotter regions with less fertile soil than maquis.

Sclerophyll woodland is dominated by oaks and is found in moister areas than either garrigue or maquis vegetation. The distribution of these three forms of vegetation in the Mediterranean Basin is shown on Fig. 9.10. The regions are largely coastal, but also extend further inland, particularly along river basins.

Sclerophyll forest is denser than woodland and is dominated by oaks and pines, largely of the same species that are more widely spaced in woodland. Forest grows in wetter locations than the other forms of vegetation (Fig. 9.10). In the north and at higher elevations, the sclerophyll tree zone merges into mixed deciduous broad-leaved and tall conifer forest.

Maquis

Maquis is the most widespread form of vegetation in the parts of the Mediterranean Basin with a mediterranean climate. It is highly varied in respect to dominant plant species and height of vegetation.

High Maquis

High maquis includes tall shrubs and small trees of varying height and favors shady slopes and sheltered locations (Fig. 4.02E). Dominant plants, such as Kermes oak (*Quercus coccifera*) and holm oak (*Q. ilex*), have dense branches and small, dark green, leathery, evergreen leaves (pages 194, 196). Strawberry tree (*Arbutus unedo*), and tree heather (*Erica arborea*) are also common. Interspersed among these are shrubs such as myrtle (*Myrtus communis*) and Spanish broom (*Genista hispanica*). Vegetation is lower-growing on the more exposed slopes and ridges.

High maquis differs from the chaparral of California in having a greater variety of plant height (Shmida and Barbour, 1981). Small trees and tall shrubs of high maquis also create a more open canopy than in typical Californian chaparral. The highest layer of trees and shrubs in maquis typically provides less than 40 percent cover, leaving more openings for low shrubs, bulbs, and annuals. In contrast, the chaparral canopy tends to occupy a single layer with more than 80 percent cover and few or no plants underneath. Compared to chaparral, maquis flora has a higher percentage of bulbs and annuals. California chaparral has a greater variety of species among its major genera: oaks (*Quercus*), California lilac (*Ceanothus*), chamise (*Adenostoma*), manzanita (*Arctostaphylos*), and buckthorn or coffeeberry (*Rhamnus*), whereas maquis has fewer species in its major genera represented by oaks, tree heather (*Erica*), strawberry tree (*Arbutus*), and myrtle (*Myrtus*).

In many parts of the Mediterranean Basin, high maquis is the form of vegetation that has developed in areas once forested by tall oak trees. When holm and Kermes oak trees are coppiced more frequently than every 20 years, they grow back not as forest, but as tall shrubs and low trees (Fig. 9.11) that are typical of maquis. However, not all maquis is thought to have resulted from wood harvesting; there is evidence that some existed in pre-

COPPICING AND POLLARDING

Coppicing. The techniques of long-term harvesting of trees by coppicing and pollarding have been used since ancient times and continue on a diminished scale to the present day. An awareness of this type of tree harvesting allows one to recognize its effects in much of the woodland and forest of the Mediterranean Basin.

Coppicing is a traditional method of woodland management that involves allowing shoots to grow from the base of felled trees (Rackham, 1995), such as holm oak. The result is a pattern of multiple trunks emerging from around a single base (Fig. 9.11). Growing shoots are harvested at intervals of two to 30 years, depending on the type of tree and the purpose for which the wood is to be used.

The first-century Roman author Columella recommended cutting oak at seven-year intervals and sweet chestnut (*Castanea sativa*) every five years. Before the age of power tools, cutting trees to a size that was appropriate to the final use represented enormous savings of labor. The new trunks were harvested when they were less than eight inches (20 cm) in diameter and were then cut to the required length. Such wood was used for fuel, fencing, light construction, and stakes for grape vines and other crops. The curved trunks near the base of a coppice were saved for building ship hulls and arched roof supports.

A major advantage of coppicing as a form of woodland management is its sustainability. Trees that would normally have a shorter lifespan instead can replace themselves by natural regrowth for more than 70 times over many centuries without showing signs of disappearance (Rackham, 1995). Unfortunately, these arts of woodmanship are becoming economically less viable, especially since the 1960s. Firewood is now the major product, and it is steadily being displaced by fossil fuels (Morandini et al., 1994). The abandonment of coppicing results in the accumulation of dead wood and increasing danger from wildfires.

Many forests still show the characteristic clusters of trunks from recent or old coppicing. They also have a characteristic group of understory plants that thrive in the light created by the periodic harvesting cycles in a manner reminiscent of the herbaceous plants that become prominent following scrubland fires. The deep shade of the pre-harvest years helps to discourage invading grasses.

Pollarding. Pollarding is a more laborious method of harvesting wood. Pollards

agricultural times (Quelzel, 1981). Examples of uncoppiced maquis exist in tall and four centuries or older stands of remote and unfarmed parts of Corsica, made up largely of tree heather and strawberry tree.

Tree Heather. Tree heather (*Erica arborea*) is in the heather family (Ericaceae). It grows as a tall feathery shrub or small evergreen tree with a distinct trunk. It has abundant,

are cut about six to 15 feet (2 to 5 m) above the base, leaving a permanent trunk called a bolling (Rackham, 1995) (Fig. 9.11). Pollarding was traditionally used to prevent grazing animals from eating the new shoots. It was most beneficial before cheap wire fencing became a more practical way of keeping animals away from the trees.

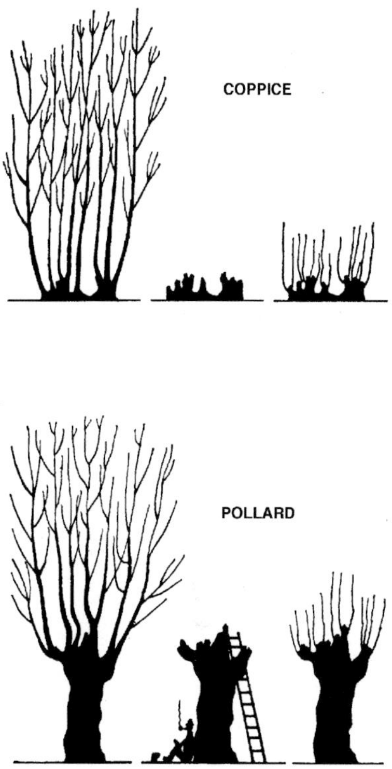

Timber. In contrast to the thin trunks that could be harvested by coppicing and pollarding, heavier timber for massive posts and beams had to be obtained from trees that were allowed to grow for as long as a century or more. European languages characteristically make a distinction between *wood* that is harvested from coppices and pollards and *timber* that is obtained from larger tree trunks.

Since the time of ancient Rome, groves of tall holm oaks were maintained to yield timber eventually, but to provide acorns to fatten hogs in the meantime. When the trees were cut, the large trunks were used for construction of houses and ships, the smaller branches for making charcoal, and the bark as a source of tannin. The Romans first named the tree ilex, which means holly.

Groves of sweet chestnut (*Castanea sativa*) have also been farmed for many centuries. Chestnuts were an important staple food in France from at least the Middle Ages. The timber has long been valued for fine furniture.

FIG. 9.11. *Coppiced and pollarded trees. Coppiced trees, illustrated in the upper panel, grow as multiple trunks from the stump of a tree whose trunk was previously cut down. These smaller multiple trunks are harvested when they reach the appropriate size for firewood, fence posts, and other purposes. New shoots emerge over the next year and gradually increase in size. Pollarded trees, shown in the the lower panel, are managed in a similar way, except that the smaller trunks are cut off the main trunk about 6 to 15 feet (2 to 5 m) above the ground. This protects the young shoots by keeping them out of the reach of grazing animals. (Reproduced with permission, Rackham, 1995.)*

fragrant, white to pinkish, bell-shaped flowers in the spring and may flower again after the beginning of autumn rains. The woody root of tree heather has long been used to make the best brier tobacco-pipes. The word brier is derived from the Old French word *bruyère* for tree heather.

Strawberry Tree. Strawberry tree (*Arbutus unedo*) is another member of the heather

A

B

C

FIG. 9.12 Maquis trees. A. The strawberry tree (Arbutus unedo) has bright fruit that turns from yellow to orange to red as it ripens in the autumn. B. and C. The Judas tree (Cercis siliquastrum) behind a chapel wall and in a closer view showing the pink blossoms and flat brown seed pods from the previous year.

family that grows as a small evergreen tree. It has clusters of cream-colored flowers late in the fall. One year later it bears globular, strawberry-like fruit with a rough exterior that turns from yellow to orange to red as it ripens (Fig. 9.12A).

Judas Tree. Judas tree (*Cercis siliquastrum*) is a de-

FOOD CROPS

More of our food crops originated in the Mediterranean Basin and the Middle East than in any of the other mediterranean climate regions. These crops include grains such as wheat and barley; legumes such as lentils, chick peas, and beans; vegetables including cabbage, asparagus, artichokes, and leeks; and fruits such as figs, grapes, olives, and almonds (Zohary, 1994). Many other plants that we now associate with the Mediterranean Basin are newcomers that were brought from other parts of the world during the age of exploration from the 15th through the 19th centuries. Among these are oranges and lemons from eastern Asia and corn, tomatoes, and peppers from the western hemisphere.

FIG. 9.13. Figs are among the many food crops that originated in the Mediterranean Basin.

ciduous member of the pea family that is native to the eastern part of the Mediterranean Basin. In spring, clusters of pink flowers are borne directly on the branches together with the brown, flattened pods that remain from the previous year (Fig. 9.12B and C). It is in the same genus as the pink-flowering western redbud (*C. occidentalis*) in California (page 75).

Mastic Tree. Mastic tree (*Pistacia lentiscus*) in the Anacardiaceae family is a particularly drought-tolerant plant of maquis. It creates mounds of dense, green foliage and is considered an excellent soil protector and builder. A variety of this tree was long cultivated for its resinous gum on the Greek island of Chios. The gum was used to keep the breath sweet, and in the 18th century, the sultan of Turkey was reported to have ordered half the annual harvest, or 125 tons, mostly for use in his seraglio (Baumann, 1993). Resin from the mastic tree was also used medicinally and as a base for varnish from ancient Roman times to the 19th century. The mastic tree is in the same genus as the pistachio tree (*Pistacia vera*), which originated further east in Persia and central Asia. The genus *Pistacia* belongs in the same family as poison oak (*Toxicodendron diversilobum*) and sugar bush (*Rhus ovata*) of California and litre (*Lithraea caustica*) of Chile.

Carob Tree. Carob tree (*Ceratonia siliqua*) is in the pea family and is a useful, drought-tolerant plant that commonly grows with the mastic tree. It is a low-growing, spreading evergreen tree that bears numerous long, brown, strap-shaped, sugary pods that take about a year to mature and ripen in autumn. The seeds are spread by animals, in whose feces they can survive and later germinate. The pods contain as much as 50 percent sugar and were used as a sweetener before sugar cane and beets became the major sources of sugar. Carob is still used as a substitute for chocolate.

Palms (Arecaceae). There are two palms native to the Mediterranean Basin. The dwarf fan palm (*Chamaerops humilis*) is a common, bushy palm growing in many of the warmer parts of the western Mediterranean Basin. In contrast, the Cretan palm (*Phoenix theophrasti*) is a rare and protected plant that is found in only a few small sites on the southern coast of Crete and in southern Turkey. The genus name *Phoenix* is appropriate to its ability to thrive after a fire. The phoenix is a mythical bird of beautiful plumage which was burned after a lifespan of many centuries only to emerge as a young bird. The Cretan palm retains its dead, dried, and highly flammable fronds on the trunk until they are burnt off by a fire. During the following rainy season, young sprouts emerge from the base of the trunk. The species name *theophrasti* is in honor of the ancient Greek botanist Theophrastus, who was a contemporary of Plato and Aristotle. The Cretan palm is in the same genus as the date palm (*P. dactylifera*), which, like most palms, is native to hotter climates.

FIRE

Fires of natural origin are less common in the Mediterranean Basin than in California, South Africa, or Australia. However, today 98 percent of the fires in the Mediterranean Basin are not natural but are caused by man (Blondel and Aronson, 1995). A high fire frequency has altered many plant communities. Plants that resprout quickly after a fire, like Kermes oak (*Quercus coccifera*), have become more dominant in large areas of southern France where fires are frequent. Most of these grow no higher than three feet (1 m). Plants with seeds that germinate after a fire, like rock rose (*Cistus* spp.), are also at a selective advantage.

After a fire in maquis, more than four-fifths of the shrub and tree species reestablish themselves by sprouting (Trabaud, 1994). Many of these have a deep taproot that provides enough water to initiate regeneration even before autumn rains begin (Margaris, 1981). In garrigue, on the other hand, where shallow-rooted species are dominant, as many recover by establishing new seedlings as by resprouting. Garrigue plants that resprout generally do not do so until the arrival of the next rainy season, because their shallow roots cannot supply sufficient water during the dry season. Seedlings emerge after the rains as well, but take longer than sprouters to establish themselves as mature plants. In garrigue, frequent fires put resprouting species at a selective disadvantage; there is less time to accumulate underground stores of the products of photosynthesis that fuel resprouting.

Plant species that can reestablish themselves only from seed are referred to as obligate seeders and include narrow-leaved cistus (*Cistus monspeliensis*), Aleppo pine (*Pinus halepensis*), and rosemary (*Rosmarinus officinalis*). In some species, the production of seedlings increases dramatically after a fire. This may be due to the heat-induced rupture of hard seed coats, allowing water to enter and the seeds to sprout. Additional benefits from fire include the inactivation of growth inhibitors in the soil and the stimulating effect of light in the red part of the color spectrum that would normally be absorbed by the leaf canopy. *Cistus* spp. can show a 20-fold increase in density of seedlings after a fire.

The response to fire in the maquis follows a time sequence similar in many respects to that described for scrub vegetation in California. In the first year after a fire, stump sprouting begins almost immediately (Trabaud, 1994). After the rains, there is an abundant growth of annuals, perennial herbs, and bulbous plants. In the second year, there is a steady growth of shrubs, annuals are prominent again, and the growth of grasses and perennial herbs reaches a peak. This is the period during which there is the greatest diversity of species. In the third and fourth years, the shrubs increase dramatically in size, and herbs are on a decline. In the fifth year and subsequently, the maquis recovers its original appearance.

A

FIG. 9.14. *Rock rose (Cistus) is a characteristic genus of maquis. A. Pink C.* crispus *is native only to the western half of the Mediterranean Basin. It is a low-growing shrub with profuse but short-lived blossoms. B.* Cytinus *is a genus of parasitic plants growing on the roots of* Cistus *shrubs.*

B

Low Maquis

Low maquis consists mainly of tall evergreen shrubs, including rock rose and oleander.

Rock Rose. Rock rose in the genus *Cistus* and family Cistaceae is among aromatic shrubs that dominate low maquis. The genus is especially common in Spain, Portugal, and Morocco. The name of the Spanish mountain range, the Sierra Morena (dark mountains), alludes to the dark green appearance of the foliage of the rock-rose-covered slopes when seen from the distance. *Cistus* species bear large numbers of five-petalled, rose-like flowers over a period of several weeks in the spring. Flower colors range from white to pink through violet and in most species last for no more than a day. Sage-leaved cistus (*Cistus salviifolius*) has white flowers and narrow leaves and *C. crispus* is pink (Fig. 9.14A). Gum cistus (*C. ladanifer*) is grown commercially in southern Portugal, Cyprus, and Crete for resin which is extracted from the leaves and used as a base for medications and perfumes (Mabberly and Placito, 1993).

Commonly associated with *Cistus* is the genus *Cytinus*, whose parasitic plants grow on the roots of *Cistus* and are found under the shrub (Fig. 9.14B). These low-growing plants have yellow, white, or pale pink flowers with conspicuous orange- or red-colored bracts.

Oleander. Oleander (*Nerium oleander*) in the Apocynaceae family is a highly poison-ous plant with a milky sap. It is generally avoided by herbivorous animals. The plant is abundant in many dry river courses often shaded by plane trees (*Platanus orientalis*) in the eastern part of the Mediterranean Basin. Oleander grows as a tall shrub and has a long summer blooming season, with pink, red, or white flowers. In modern Greece, oleander leaves are still used to stuff into mouse holes. The mice reportedly die after they nibble at the leaves to get out (Baumann, 1993).

Garrigue

The low shrubs of the garrigue are dotted over hillsides with intervening bare, stony or sandy patches (Fig. 9.15B). Garrigue is a widely used French term. Phrygana is a Greek word and is commonly used for similar or even lower-growing forms of vegeta-

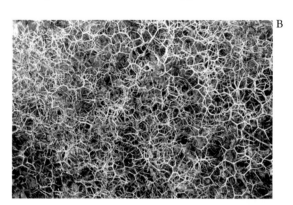

B

FIG. 9.15. *Garrigue. A. A landscape in southern Crete that is dominated by low-growing cushion-shaped shrubs on rocky soil. The gold-green mounds are Greek spiny spurge* (Euphorbia acanthothamnos). *B. Most of the gray-green shrubs are thorny burnet* (Sarcopoterium spinosum), *which has a hexagonal skeleton of thorny spines.*

A

tion in the eastern Mediterranean. Batha is the corresponding term in Israel. Garrigue and its equivalents are found in the hottest and driest parts of the mediterranean climate. In Israel, for example, it accounts for more than 40 percent of all hilly, upland terrain (Shmida and Barbour, 1982). Much of the garrigue has been heavily grazed for thousands of years.

Spiny Cushion Plants

Many of the most prominent plant species of the garrigue are spiny plants that are not attractive to grazing animals. They often form distinct cushion shapes, leaving room for bulbs and orchids in the intervening spaces. Among these cushion plants is Greek spiny spurge (*Euphorbia acanthothamnos*), which forms gold-green mounds in the spring. It is often found growing with thorny burnet (*Sarcopoterium spinosum*) (Fig. 9.15A), another mounding plant that has a thorny skeleton with a hexagonal pattern (Fig. 9.15B). Both species are native to the eastern part of the Mediterranean Basin.

Drought-Deciduous and Dimorphic Plants

As in the coastal sage scrub of California, many garrigue plants produce two types of foliage and are referred to as seasonally dimorphic (page 33). Jerusalem sage, for example, has relatively large and soft leaves, which emerge in the winter rainy season. Later, in spring and early summer, a more drought-tolerant foliage is produced, and winter leaves are lost. Other plants, such as tree spurge (*Euphorbia dendroides*) and spiny broom (*Calycotome villosa*), are completely drought-deciduous. Spiny broom is member of the pea family and is most conspicuous in the spring when it has clusters of yellow flowers. It releases its seeds in July and sheds its dried leaves in August. With the autumn rain, leaves reappear on the new branches that were produced the previous spring.

Aromatic Plants of the Mint Family (Lamiaceae)

The aromatic plants in the garrigue correspond in many ways to those of the coastal sage scrub or soft chaparral of California (Shmida and Barbour, 1982). In both plant communities, there are abundant shrubs that emit oils with a pungent odor, particularly in the heat of day. Plants in the mint family are particularly common in the Mediterranean Basin. Examples include rosemary (*Rosmarinus officinalis*) and Spanish lavender (*Lavandula stoechas*), both of which have blue to violet flowers. Rosemary grows best near the sea, hence its name *ros marinus*, meaning dew of the sea. It is widely used for cooking, particularly in France and Italy, and its oil is used for perfume. French lavender (*L. dentata*) is used to give its fragrance to linens and strewn on the floor to disguise unpleasant odors. It is also used in perfumes. Marjoram and oregano are among the other aromatic plants of the mint family that are used in cooking. Three-leaved sage (*Salvia triloba*), another of the mint family, is used to make a herbal tea called fascomelo in Greek. The flowers are pale pink-lilac or white. Among the most attractive flowers of the sages are those of Jerusalem sage (*Phlomis fruticosa*), whose bright yellow blossoms form in dense whorls, one above the other

FIG. 9.16. *The lily family (Liliaceae). A., B. A red tulip endemic to Crete,* Tulipa doerfleri, *favors previously cultivated fields. C. The pink rock tulip (*T. saxatilis*) is found in both Crete and southwestern Turkey. D. Yellow asphodel (*Asphodeline lutea*) often grows with E. the white-flowering common asphodel (*Asphodelus aestivus*), which is native to the eastern Mediterranean Basin. Both are unpalatable to sheep and goats.*

on the stems (Fig. 3.05B). The oils of many of these aromatic plants of the mint family have long appealed to man, but discourage consumption by foraging animals.

Bulbs

After the Western Cape, the Mediterranean Basin has the greatest diversity of bulbs and other geophytes. These are most abundant in the garrigue, where narcissus, crocus,

D

E

tulip, hyacinth, and iris grow among the shrubs. Peak bloom is in spring. In summer and early fall, the landcape becomes rather dry and harsh. But it turns green again soon after the first autumn rains, and there is a second bloom, which includes some species of narcissus and crocus.

Lily Family. Tulips (*Tulipa*) are in the lily family (Liliaceae) and grow from the eastern part of the Mediterranean Basin (Fig. 9.16A to C) to the Central Asian grasslands. Tulips were propagated in Turkey in the 16th century. They were brought to Holland in the 17th century, where they commanded enormous prices during a brief tulip craze.

Also in the lily family are asphodels, which have fibrous or tuberous roots. They are abundant in heavily grazed areas on rocky hillsides because they are typically avoided by animals. The common asphodel (*Asphodelus aestivus*) with its tall stalks of white flowers is often found growing with the yellow asphodel (*Asphodeline lutea*) (Fig. 9.16E and D).

Amaryllis Family. The genus *Narcissus* belongs to the amaryllis family (Amaryllidaceae). It is found in most parts of the Mediterranean Basin, but the greatest number of species is concentrated in the west, in Spain, Portugal, and Morocco. Most species are found in hills, mountains, and along valleys, where they favor a northern exposure. The common hoop petticoat narcissus (*Narcissus bulbocodium*) has a yellow, solitary flower on each stem and is native to the Iberian Peninsula, southwestern France, and northwestern Africa. It was grown in England in 1789,

FIG. 9.17. *The common hoop petticoat narcissus,* Narcissus bulbocodium, *is on the left. It has a yellow, solitary flower and is native to the Iberian Peninsula, southwestern France, and northwestern Africa. The illustration appeared in* Curtis's Botanical Magazine *in 1789. On the right is a slightly later print of the polyanthus or rose of Sharon narcissus (*Narcissus tazetta*), which has clusters of two to seven white flowers with deep yellow cups. It is found in much of the Mediterranean Basin but its distribution also extends much further to the east, to China.*

FIG. 9.18. *The iris family (Iridaceae).* Crocus biflorus, *on the left, is a spring-blooming species found in Italy and Turkey. Its fragrant flowers range from white to lilac-blue.* Iris florentina *is a tall, white, blue-veined iris with a pale yellow beard that has long been cultivated in the Mediterranean Basin. Both prints appeared in* Curtis's Botanical Magazine *at the turn of the 18th to the 19th century.*

when the print shown in Fig. 9.17 appeared in the Botanical Magazine. The polyanthus narcissus or rose of Sharon narcissus (*Narcissus tazetta*), shown in the same figure, has clusters of two to seven white flowers with deep yellow cups. It has a similar natural distribution and is also widely cultivated. Narcissus bulbs contain a toxic narcotic. The ancient Greeks associated the plants with death and put narcissus wreaths in the hands of the deceased (Huxley and Taylor, 1989).

Iris Family. Plants of the iris family (Iridaceae) are found on all continents except Antarctica, but the genus *Iris* is native to the northern hemisphere. Wild forms of the bearded iris (Fig. 9.18), the most popular garden iris, grow in the Mediterranean Basin and adjacent parts of central Europe and central Asia.

The genus *Crocus*, also in the iris family, grows throughout most of the Mediterranean Basin and into the Middle East and central Asia, extending as far as western China. Most of the 80 or so species grow in Turkey, Greece, and the Balkan peninsula. In contrast, there are only four species in the Iberian Peninsula and three in the central Asian republics of the former Soviet Union (Mathew, 1982). *Crocus* corms can be edible, and in Greece, some species are roasted and are known as kastanea, the same word that is used for chestnut. *Crocus sativus* is a widely cultivated lilac-colored species that is the source of saffron, which comes from its laboriously collected styles. *Crocus biflorus* (Fig. 9.18) is a spring-blooming species found in Italy and Turkey. Its fragrant flowers range from white to lilac-blue.

Cyclamen. *Cyclamen* is a genus in the primrose family (Primulaceae), found mainly in the eastern part of the Mediterranean Basin, in Italy, Greece, the Balkans, and Turkey. It grows from underground corms. *Cyclamen persicum* is the most commonly grown species and is the ancestor of the florists' cyclamen. In the wild, it blooms from January to May. Several other *Cyclamen* species flower in the fall, after the first rains. The common English name for *Cyclamen*, sowbread, is based on its tubers being a favorite food of wild boars.

Orchids. The orchid family Orchidaceae is a very large one. It is well represented in the Mediterranean Basin with numerous species that are all terrestrial. This is in contrast to the epiphytic, tropical species that grow on the branches and trunks of trees. Mediterranean orchids are particularly common in rocky garrigue that is used for grazing. Some species are so abundant that they are seen as cut flowers in inns and restaurants to the dismay of conservation-minded guests. Other orchids are so rare that positive identification by amateur field-botanists is an occasion for euphoria and celebration.

A selection of orchids from the highland plateaus of central to western Crete is shown in Fig. 9.19. This region is one of the prime destinations for wildflower trips in the Mediterranean Basin. The Italian man orchid (*Orchis italica*) is also common in many other parts of the Mediterranean Basin. Its pink to lilac flowers are borne in oval clusters, each one appearing to have a human male shape with slim, pointed limbs and a penis. The deep yellow *Orchis pauciflora* is also relatively common and widely distributed in the Mediterranean Basin as is *Orchis boryi* with its pink, violet, and purple flowers. These species often grow together in clusters.

The Mediterranean Basin is notable for its numerous species of the orchid genus *Ophrys*, whose flowers have a distinctly insect-like appearance, in many species resembling a bee. The flowers may also produce the sexual odor of a female insect. These characteristics attract the males of specific insect species which then attempt to copulate with the flower, picking up pollen in the process. The insect, not having learned from its mistake, moves on to another "female," resulting in the pollination of another flower.

A

B

C

D

FIG. 9.19. Orchids. A. and C. The Italian man orchid
(Orchis italica) has pink to lilac flowers, borne in oval
clusters, each one having a human male shape. B. Orchis
boryi has pink, violet, and purple flowers, often all growing
together in clusters, as illustrated. D. The sawfly orchid
(Ophrys tenthredinifera) represents the genus Ophrys,
whose flowers are not only characterized by a distinctly
insect-like appearance, but may also produce the sexual
odor of the female insect. E. The small-flowered tongue
orchid (Serapias parviflora) belongs to another genus that
is associated primarily with the Mediterranean Basin.

E

The life cycle of orchids seems unusually beset by perils. Extremely light seeds are wind-dispersed; they must land in an area of disturbed soil and meet the appropriate fungus for an essential mycorrhizal association. In order to flower, young plants must remain relatively undisturbed for 15 to 20 years, typically in areas that are used for grazing. With these circumstances, it is surprising that there are places where orchids are the most abundant flowering plant.

Woodland and Forest

Woodlands and forests of the Mediterranean Basin are dominated by oaks and to a lesser degree by conifers. Forests now cover less than five percent of the area that was forested in prehistoric times. Table 9.01 shows the current extent of forest in countries

TABLE 9.01. FOREST, MAQUIS, AND GARRIGUE IN THE MEDITERRANEAN BASIN

Countries that have a mediterranean climate and that border the Mediterranean Sea are listed in clockwise order, starting with Spain. Countries with more than 1 million hectares (2.5 million acres) of mediterranean vegetation are shown in bold. (Data from Grenon and Batisse, 1989.)

	Total Forest, Maquis and Garrigue hectares, millions	Mediterranean-type Component hectares, millions	Mediterranean-type Component percent of total
Spain	26.8	9.1	34
France	14.6	2.2	15
Italy	6.1	1.5	25
Yugoslavia (former)	9.1	0.9	10
Albania	1.3	0.3	20
Greece	2.6	1.6	60
Cyprus	0.2	0.2	100
Turkey	12.3	6.1	50
Syria	0.4	0.4	100
Lebanon	0.1	0.1	100
Israel	0.1	0.1	100
Libya	0.5	0.5	100
Tunisia	0.8	0.8	100
Algeria	2.4	2.4	100
Morocco	5.2	5.2	100

PROLONGED IMPACT OF MAN

The landscape and vegetation of the Mediterranean Basin have been profoundly influenced by man for many thousands of years. Egypt, the Middle East, and Turkey were among the earliest places settled by agricultural societies. Cultivation of soil promoted a spread of annual weeds. The grazing of livestock favored the survival and proliferation of pungent aromatic herbs and thorny or poisonous shrubs that are avoided by grazing animals (Fig. 9.20).

Interestingly, a moderate degree of grazing may favor species diversity in some of the areas that are among the most popular destinations for wildflower enthusiasts. Naveh and Whittaker (1979) describe a fourfold reduction in the number of plant species in northern Israel when grazing was completely stopped in an area of open shrublands, allowing the return to a closed canopy of maquis shrubs. The authors feel that moderate grazing is necessary to maintain the abundant spring flowering in these landscapes.

Deforestation of the Mediterranean Basin has had many negative consequences. With the growth of urban societies, the demand for wood as fuel and construction material has steadily increased and changed the landscape. The geography of the Mediterranean Basin favored sea transport of resources from one region to another. Virtually treeless Egypt, for example, continuously imported wood, gradually denuding the once extensive forests of Crete, Cyprus, and Lebanon.

Deforestation occurred for a far longer time in the Mediterranean Basin than in any other mediterranean climate region. Plato, Pliny, and other ancient authors warned about the serious effects of deforestation (Baumann, 1993). When the Persian ruler Xerxes invaded Greece in 480 B.C., his army constructed a wooden pontoon bridge across the narrowest part of the Dardanelles strait between Turkey and the Balkan peninsula. It consisted of an estimated 21,000 tree trunks fastened across 674 galleys of 50 oars each, anchored side by side. Nearby forests must have been denuded to a large extent during this phase of the campaign.

Durable wood of Italian cypress (*Cupressus sempervirens*) was exported from Crete for thousands of years. In 1414, when Crete was a colony of Venice, the Venetian Senate issued a decree forbidding the further export of cypress wood. However, continued harvesting of wood has made Crete an island with few patches of woodland and forest, today mostly restricted to deep ravines and steep mountain slopes.

As the Phoenician, Greek, and Roman empires spread, they created active trading routes. Most present-day Mediterranean towns date at least from Roman times. They have expanded since but maintain their original locations. Terraces of wheat, grapes, and olives were widespread by Roman times and were abandoned only recently as they became unsuitable for large-scale, mechanized farming.

Steep mountains jutting down to the sea often made land travel more difficult than voyages in many parts of the Mediterranean Basin. Rugged topography fostered development of many culturally distinct areas, currently comprising more than 18 countries, each with its own distinctive cultures and languages. Most of these nations and their provinces were surprisingly self-sufficient until the growth of railroads and steam shipping made transport relatively inexpensive in the mid-19th century. During the present century, a marked increase in population, the growth of industry and transport, and the explosive rise in tourism have had major environmental effects (Fig. 9.20).

At present, prospects for forest preservation are best in Spain, France, and Italy, where wooded and forested areas are actually increasing as many economically marginal farms are abandoned and where over 70 percent of the population is urban. Population growth is slow in these countries, and the remaining native trees are no longer suitable for structural timber. Most lumber is imported from more temperate or tropical regions. Even the harvesting of firewood does not remain economically competitive with other forms of fuel.

The prospects for protecting woodland and forest are less favorable along the eastern and southern shores of the Mediterranean, where a less prosperous population is expanding rapidly with over half remaining in rural areas. Increasing population pressure, grazing, fuel gathering, and clearing of marginal land for agriculture lead to loss of trees and growing areas of irreversible desertification.

F𝗂𝗀. 9.20. *Appearance of a Mediterranean coastal slope before the advent of railroads and steam ships in the 19th century (top panel). Villages were largely on the hillsides, away from the malarial coastal swamps and lagoons. They were virtually self-sufficient, raising varied crops on laboriously terraced land and moving flocks of grazing animals from lowland winter pastures to summer pastures in the higher mountains. Forest, woodland, and maquis were harvested for fuel and building materials. (From Grenon and Batisse, 1989, with permission of Oxford University Press.)*

facing the Mediterranean Sea (Grenon and Batisse, 1989). By far the most extensive areas of sclerophyll forest occur in Spain, Turkey, and Morocco. Substantial woodlands and forests also grow in France, Italy, and Algeria.

Oaks

Holm oak. Holm oak (*Quercus ilex*) forest once covered much of the Mediterranean Basin. Holm oak now grows throughout the mediterranean area, but is most abundant in the western half of the region. The tree is drought-resistant, and the density of its canopy tends to keep soil underneath moist. Holm oak is found from sea level to an elevation of 6,500 feet (2,000 m). Along the coast, it tolerates salt winds better than other species of oak.

FIG. *9.21. Oaks. The dark foliage of the evergreen holm oak* (Quercus ilex) *and Kermes oak* (Q. coccifera) *line both sides of the Patmos Gorge in Crete. The pale green foliage of plane trees* (Platanus orientalis) *flanks the stream.*

The largest forests and woodlands of holm oak are found on the Iberian Peninsula, where it often hybridizes with cork oak (*Q. suber*). Extensive oak woodlands are used for grazing (Serrada et al., 1995). Much of the wooded and forested area is covered by coppiced holm oak, which is harvested to supply firewood for weekend and holiday residences. Holm

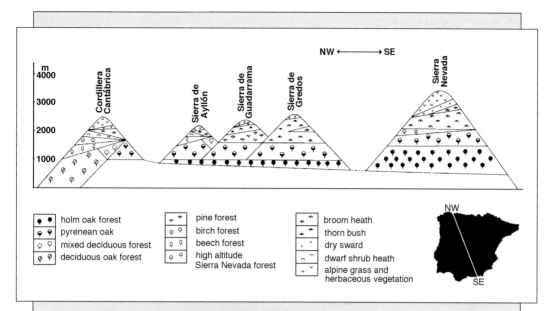

FIG. 9.22. *Effect of elevation and geographical position on vegetation. A cut through Spain illustrates the transition from the temperate, year-round rainfall area near the Atlantic coast in the northwest to the mediterranean climate area of the central plateau and Mediterranean coastline in the southeast. (From Ern, 1966.)*

ELEVATION

The effect of elevation on the vegetation is profound, as it is in California and Chile. This is illustrated in the diagram (Fig. 9.22) showing a cut through Spain from the temperate, year-round rainfall area near the Atlantic coast in the northwest to the mediterranean climate area of the central plateau and Mediterranean coastline in the southeast (Ern, 1966). In a distance of about 525 miles (850 km), five mountain ranges are crossed, culminating in the highest range, the Sierra Nevada. Toward the southeast comparable plant communities extend to higher elevations, partly due to the warmer latitudes. The evergreen, sclerophyll holm oak (*Quercus ilex*) is common on all but the northernmost range. In the central three ranges it is found between about 2,600 and 3,300 feet (800 and 1,000 m). In the Sierra Nevada, it occupies a relatively broad band, extending to 5,000 feet (1,500 m). Similarly, the next higher elevation is dominated by the deciduous pyrenean oak (*Q. pyrenaica*), which is found up to about 5,000 feet (1,500 m) on the northern four ranges and extends to 6,600 feet (2,000) in the southern slope of the Sierra Nevada. Above that elevation, the Sierra Nevada is covered mainly by broom heath, thorn bush, herbaceous vegetation, and grasses. The wet, Atlantic-facing slope of the northernmost Cordillera Cantábrica is well out of the mediterranean climate region and has deciduous oak, beech, and birch forest more typical of central and northern Europe.

oak is rarely found growing to its full potential height of 80 feet (25 m). There are still ancient forests of holm oak in Sardinia, with trees that are up to 1,000 years old. These forests are grazed by livestock, thereby reducing the danger of fire from accumulation of understory brush (Naveh, 1982).

Holm oak yields dark brown timber that is hard, heavy, and durable but difficult to work. Today the tree is more valued as an ornamental plant for its abundant, dark evergreen foliage (Fig. 9.21) and corrugated trunk. Many fine old specimens planted in the mid-19th century are now a prominent feature of the Royal Botanic Gardens at Kew in England. Holm oak is described as "in many respects the finest of all evergreen trees, apart from conifers, cultivated in the British Isles" (Bean, 1976).

Holm oak has leathery corrugated leaves that are green on top and have gray felted hairs underneath. Leaves are highly variable in shape, some having spiny margins like holly leaves and others having smooth edges. As with many other broadleaved evergreen trees, leaves last two to three years. Acorns ripen in a single year.

Kermes Oak. Kermes oak (*Quercus coccifera*) is another widespread evergreen oak. It usually grows as impenetrable, holly-like bushes and low trees scattered over hotter and drier hillsides. Because the Kermes oak tolerates harsh conditions, it sometimes colonizes land previously dominated by holm oak, particularly following a forest fire. In the eastern Mediterranean, it is supplanted by the closely related Palestine oak (*Q. calliprinos*), which some consider to be a subspecies of the Kermes oak. The Palestine oak can be found growing to a height of 65 feet (20 m) in cemeteries and sacred groves, where it has been left undisturbed for hundreds of years (Naveh, 1994). Natural forests of Palestine oak remain only in small areas of Greece, in Crete and the Peloponnese.

After extensive damage by fire or harvesting of wood, Kermes and Palestine oaks regenerate from basal sprouts to form very dense brush that shelters herbaceous plants beneath. This low form is widespread and is estimated to cover more than five million acres (two million hectares) in the Mediterranean Basin. The two species can be distinguished from holm oak by the lack of a velvety underside of their shiny, dark green, spiny leaves. Also, the acorns, which take two years to ripen, have cups with rigid spiny scales that radiate to all sides, whereas those of holm oak have scales that adhere closely to the cup.

The Kermes oak derives its name from the Kermes scale insect (*Coccus ilicis*) that breeds on it. The insect was the source of one of the most prized dyes of the ancient and medieval worlds, the scarlet that is mentioned in the Bible and was called "Grain of Portugal" by Chaucer (Polunin and Huxley, 1990). Although the Spartans did not allow the use of dyes for ordinary clothing, the red dye was used for garments worn in battle so that bloodstains would be less conspicuous (Baumann, 1993).

Cork Oak. Cork oak (*Quercus suber*) is most common in southern Portugal, southwestern Spain, and Morocco where it is most often seen in orchard-like groves maintained by some clearing of understory brush. The tree grows to 50 feet (15 m). Its thick, deeply fissured, spongy bark is stripped for cork about every 10 years, without endangering the tree. The freshly stripped trunk has a red appearance and gradually regenerates new bark from the

A

B

FIG. 9.23. A. Downy oak (Quercus pubescens) is a common deciduous oak conspicuous for its picturesque twisting branches. B. The large fringed acorn cup is a characteristic feature of the deciduous valonia oak (Q. macrolepis), which grows over a broad range in the southern Balkans.

dividing cells of the cambium layer, which is left intact. Woodlands of cork oak have been harvested for many centuries, particularly in Portugal and Spain. Like old olive groves, they have become picturesque with their open foliage and angular branching pattern.

In contrast to the holm, Kermes, and cork oaks, which are all evergreen trees with tough, sclerophyllous leaves, the downy oak (Q. pubescens) and valonia oak (Q. macrolepis) are deciduous.

Downy Oak. Downy oak (Quercus pubescens) is the most common deciduous oak in the Iberian Peninsula, southern France, and Italy. Its gray-green leaves are lobed and downy beneath. From the distance, downy oaks (Fig. 9.23A) often have a silhouette that is similar to certain California oaks, such as valley oak and blue oak. In the spring, the new light

green foliage of the deciduous oaks forms a sharp contrast with the dark green of the evergreen oaks.

Valonia Oak. Valonia oak (*Quercus macrolepis*) is one of several species of deciduous oaks in the northeastern part of the Mediterranean Basin. It has huge, distinctive "mossy" acorn cups (Fig. 9.23B). The deeply divided leaves are a shiny dark green above and gray-green and velvety below.

Conifers

There are three Mediterranean pines that commonly grow close to the sea. All three have needles in pairs and often have an umbrella-like crown.

Stone or Umbrella Pine. Stone or umbrella pine (*Pinus pinea*) develops into an umbrella shape most consistently, growing to 80 feet (25 m) (Fig. 9.24). It has been planted since ancient times for its edible seeds, called pignole, pignon, or pine nuts. The seeds are borne on large, shiny, red-brown cones that take three years to mature, ripening in early spring

FIG. 9.24. *Stone or umbrella pine* (Pinus pinea). *(Drawing reproduced with permission, Farjon, 1984.)*

and opening to release seeds on very hot summer days or after a fire. Nuts are harvested by pulling the closed cones off the trees with a long, hooked pole. The cones open to release their nuts when spread out in the sun or heated in an oven. There can be over 100 nuts in one cone.

Forests of stone pine are found on the Iberian Peninsula, Italy, Greece, and Turkey, but it is hard to say which groves are natural, because the tree has been so extensively planted. One of the most famous forests known to have been planted is on the Adriatic coast near the Italian city of Ravenna. The forest is 16 miles long and one mile wide (1.6 by 25 km). It has been in existence for over 1,500 years and is mentioned in Dante's *Purgatory* (Bean, 1976).

Maritime Pine. Maritime pine (*Pinus pinaster*) grows primarily in the western Medi-

FIG. 9.25. *Maritime pine (Pinus pinaster). (Drawing reproduced with permission, Farjon, 1984.)*

terranean, from Morocco in the south to Spain and France to Italy in the north. Its branches extend from the top third of a reddish trunk that reaches a height of 100 feet (30 m). The maritime pine has dark foliage with long, sharp needles (Fig. 9.25). It can form dense woods with little undergrowth, or grow as open woods with an understory of evergreen plants of the maquis, such as tree heather, strawberry tree, and rock rose. It is more frost-sensitive than the stone pine. Since the 16th century, the tree has been Europe's primary source of turpentine. More recently, plantings have been used to stabilize dunes.

Italian Cypress. Italian cypress (*Cupressus sempervirens*) grows in two distinct shapes, thin and columnar (Fig. 9.26) or more irregular and horizontal. Both are na-

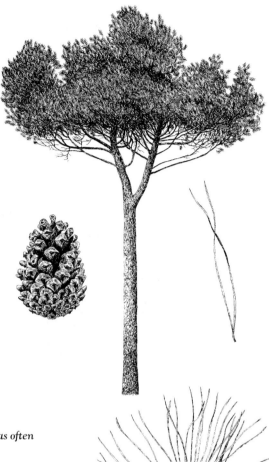

FIG. **9.26.** *Italian cypress (*Cupressus sempervirens*) has often been planted around chapels or cemeteries.*

FIG. **9.27.** *Aleppo pine (Pinus halepensis). (Drawing reproduced with permission, Farjon, 1984.)*

tive to Greece, but are also frequently planted in the Mediterranean Basin. Although it was a timber tree in ancient times, more recently it is largely used as an ornamental. In Greece, it grows around cemeteries and chapels.

Aleppo Pine. Aleppo pine (*Pinus halepensis*) is found along the hotter parts of the Mediterranean coast. It is the most drought tolerant and most susceptible to fire of these three pines. The tree typically has a round or pyramidal crown (Fig. 9.27). It is the most common and most widely distributed pine near the coasts and is usually found in groves of scattered trees mixed with the maquis or garrigue type of vegetation of lavender, rosemary, thyme, and rock rose growing in intervening open places. Aleppo pine is rich in resin, which in Greece is used to flavor wines called *retsina*.

PLANTS FOUND IN MEDITERRANEAN CLIMATE AREAS OF THE MEDITERRANEAN BASIN

Arbutus unedo	strawberry tree	Ericaceae	small evergreen tree
Castanea sativa	sweet chestnut	Fagaceae	deciduous tree
Ceratonia siliqua	carob tree	Fabaceae	evergreen tree
Chamaerops humilis	dwarf fan palm	Arecaceae	low-growing palm
Cistus monspeliensis	rock rose	Cistaceae	evergreen shrub
Crocus spp.	crocus	Iridaceae	bulbous
Cyclamen spp.	cyclamen	Primulaceae	bulbous
Erica arborea	tree heather	Ericaceae	evergreen tree/shrub
Iris spp.	iris	Iridaceae	bulbous
Juniperus oxycedrus	prickly juniper	Cupressaceae	evergreen shrub/tree
Lavandula stoechas	Spanish lavender	Lamiaceae	evergreen shrub
Myrtus communis	myrtle	Myrtaceae	evergreen shrub
Narcissus spp.	daffodil, narcissus	Amaryllidaceae	bulbous
Nerium oleander	oleander	Apocynaceae	evergreen shrub
Olea europaea	olive	Oleaceae	evergreen shrub/tree
Pinus halepensis	Aleppo pine	Pinaceae	evergreen conifer
Pinus pinaster	maritime pine	Pinaceae	evergreen conifer
Pinus pinea	stone or umbrella pine	Pinaceae	evergreen conifer
Pistacia lentiscus	mastic tree	Anacardiaceae	evergreen shrub/tree
Quercus coccifera	Kermes oak	Fagaceae	evergreen tree/shrub
Quercus ilex	holm or holly oak	Fagaceae	evergreen tree/shrub
Quercus pubescens	downy oak	Fagaceae	deciduous tree
Quercus suber	cork oak	Fagaceae	evergreen tree
Rosmarinus officinalis	rosemary	Lamiaceae	evergreen shrub
Sarcopoterium spinosum	thorny burnet	Rosaceae	shrub
Tulipa spp.	tulip	Liliaceae	bulbous

A

B

C

D

E

F

G

10.
PLANNING A TRIP

N atural history trips are becoming increasingly appealing, especially for amateur naturalists who appreciate outdoor activities. There is an ever greater variety of tours being offered with an emphasis on wildflowers, woodlands, and forests. Many are led by a well informed naturalist, the pace is relaxed, without strenuous hiking. Such tours are typically sponsored by museums, botanical gardens, and specialized travel agencies. They have the advantage of taking care of all the details involved in planning the trip and of providing expert and often scholarly information throughout. Overseas tours originating in the United States typically cost about $4,000 to $8,000 (1998 values) for a trip of two to three weeks. They may include attractions not available to the general public and provide an opportunity to make new friends among a small group with similar interests.

It is often less costly to book a tour that is locally organized and accompanied by a local naturalist guide, as can be done through travel agencies in Perth, Australia, or Cape Town, South Africa. Similarly, there is a great variety of natural history tours to the Mediterranean Basin and other destinations originating in Great Britain (page 228). These options require faxing, telephoning, or writing for information and doing more advance planning.

For the experienced traveller, making one's own arrangements, possibly with the help of a good travel agent, will have the advantages of flexibility. There are now brochures and books that help to plan an independent trip to almost all of the mediterranean climate areas. A winner in this category is the free booklet from the Western Australia Tourist Centres, entitled *Wildflower Discovery: A Guide for the Motorist* (page 221). A very helpful book focused on natural history travel in the Mediterranean Basin is *Mediterranean Wildlife* by Raine (1994). Especially useful books in the Lonely Planet series are those on South Africa, Australia, Western Australia, and Chile.

Some of the books recommended in this chapter are not easy to find, even in large bookstores. However, most can be obtained by special order. There are mail order services specializing in books on natural history. These include Flora and Fauna Books, 121 First Ave. South, Seattle, WA, Tel. (206) 623-4727 and Fax (206) 623-2001. In the United

FIG. 10.01. A-C. *South Australia. Kangaroo Island wildlife: A. The unusual echidna, a spiny mammal that lays eggs, like the platypus; B. Koala bear; C. Seals. D. Crete. Donkey grazing. E. Crete. Arkádhi Monastery. F. Crete. Village in the Amari Valley. G. Western Australia. The Plantagenet Hotel in Mt. Barker, renowned for its steak.*

Kingdom, Natural History Book Service Ltd., 2-3 Wills Road, Devon TQ9 5XN carries a large list of titles and issues a catalog twice a year (Tel 44-1803-865913 or Fax 44-1803-865280). Internet users will find an enormous variety offered by http://www.amazon.com.

The remoteness of most mediterranean climate areas results in relatively high travel expenses, even using APEX fares and ticket discounters. Most visitors to Chile, South Africa, and Western Australia travel a long way and may not return again. This leads to the temptation to pack too many attractions into a brief period in a vain effort to avoid missing anything. It is worth making some difficult choices to slow the pace, allowing the trip to be more relaxing and enjoyable.

One of the important advantages of nature travel to mediterranean climate destinations is that they are at their best outside peak travel times, from March through May in the northern hemisphere and from August through October in the southern hemisphere. Planes and airports do not have holiday crowds, off-season fares are available, and it is relatively easy to reserve campsites and accommodations and to rent cars or bicycles.

All of the mediterranean climate areas have additional attractions that add to the pleasures of a trip. Without exception, they have wine growing regions, are often noted for their pleasant towns, informal accommodations, and interesting restaurants. All have beautiful beaches for walks and picnics that are relatively empty outside the summer season. In California, Chile, and parts of the Mediterranean Basin (especially Spain, France, and Italy), a wildflower trip can easily be combined with late-season skiing. Orange groves and vineyards are within view of snowcapped mountains. A trip to South Africa would seem incomplete without a visit to one of the big game reserves that are only a short flight away. The flight to Western Australia is likely to require a stop in Sydney, one of the world's most beautiful cities, or perhaps Brisbane or Cairns for a detour to the Great Barrier Reef, another great nature travel destination. The Mediterranean Basin occupies such a large and varied area that it is difficult to visit more than a very small part of it. What this area offers to compensate for its relative paucity of unspoiled natural sites is a rich cultural setting which allows one to stay in beautifully situated historic towns, giving an additional dimension to the trip.

Travel information is apt to change from one year to the next. Where addresses and phone numbers are listed, check whether they are still current.

A. California

The best period for wildflowers in the mediterranean climate area of California is from March through May. However, the state's large forests and woodlands and the mild climate along the coastal strip make for a pleasant trip at almost any season.

For nature travellers, there are many books describing hiking trails primarily in national and state parks and in national forests. Others describe country roads and bed-and-breakfast destinations. These books, though helpful in planning a trip, generally give little information about the natural attractions of the state, especially its plants.

Perhaps the best way to start planning a nature trip is to choose among the many excellent books that describe the diverse plant communities of the state. There are several interesting and readable examples, the most compact of which is the popular *Introduction to California Plant Life* by Ornduff (1974) still being reprinted. A somewhat different perspective is presented in *An Island Called California: An Ecological Introduction to its Natural Communities* by Bakker (1984) and more recently in *California's Changing Landscapes. Diversity and Conservation of California Vegetation* by Barbour et al. (1993). All three books explain the different climates, soils, and plant communities of the state and mention some of the outstanding travel destinations. A more comprehensive guide to the natural history of the state is provided by Schoenherr's *Natural History of California* (1992). Some of the interesting ecologic interactions in woodland and forest are covered in Johnston's *California Forests and Woodlands* (1994).

A book that successfully combines interesting and non-technical botanical information with descriptions of landscape and travel is *Oaks of California* by Pavlik et al. (1991). The book is beautifully illustrated and has a large section entitled "Exploring California's Oak Landscapes," which outlines hikes and scenic drives in various parts of the state where oak woodlands can be enjoyed, often in combination with forest, chaparral, and grasslands. Detailed driving instructions and phone numbers to call for information are provided.

There is a third category of books that deals with the natural history of specific regions. Two excellent examples are Henson and Usner's, *The Natural History of Big Sur* (1993) and *The Natural History of the Point Reyes Peninsula* by Evens (1993). The first concentrates on the thinly populated and scenic central coastline of California. The last third of the book describes various hiking trails in this region with detailed instructions on how to reach the trail heads and what forms of vegetation and plant species are to be seen. The second book gives a beautiful overview of the National Seashore area just north of San Francisco.

Visitors to California who arrive by air generally land in San Francisco or Los Angeles, the two largest metropolitan areas. Car rental is almost essential. Fortunately, the costs of rental and fuel are lower than in any of the other mediterranean climate areas.

Offices of the California State Automobile Association (CSAA) can be found in many locations. These offer free maps, booklets, and touring information not only to its own members, but also to those presenting membership cards of automobile clubs from other countries. The staff is usually well-informed, especially about common destinations such as the Big Sur, the redwood parks, the gold country, and Yosemite National Park. Detailed county maps of areas to be visited are particularly useful.

SAN FRANCISCO AREA

In the San Francisco area, there are several botanical gardens where native plants are well labeled and one can get a good introduction to California's flora. The **Strybing Arboretum and Botanical Gardens** is in San Francisco's Golden Gate Park. In addition to a section for California native plants, it has gardens to represent other mediterranean regions. Across the Bay, there are two other botanical gardens of interest in Berkeley. The **University of California Botanical Garden** also has plants arranged according to their geographical origin in a beautiful hillside setting with views of San Francisco Bay. Especially noteworthy is the **Regional Parks Botanic Garden** in **Tilden Regional Park** at the top of the Berkeley hills. It specializes entirely in native plants, which are arranged according to the different climates that are found in the state.

Within a 90-minute drive of San Francisco, there are numerous places to enjoy all of the major plant communities that have been discussed. The choice depends mainly on what one is planning to do for the rest of the trip. North of San Francisco, in Marin County, the **Point Reyes National Seashore** is one of the most beautiful destinations. The excellent visitors center on Bear Valley Road has a small bookstore and information on a variety of trails that feature oaks, chaparral, and coastal vegetation. **Muir Woods National Monument** is the closest natural redwood forest to San Francisco. It is less than a 30-minute drive from the Golden Gate Bridge on a mountain road that lies within **Mount Tamalpais State Park.** The latter has oaks, chaparral, and grassy meadows full of wildflowers in the spring. There are many trails with dramatic views of the ocean and city.

A little more than an hour south of San Francisco via Highway 280 are a number of interesting state parks. Among these is **Big Basin Redwoods State Park**, established in 1902 and the oldest park in the state system. It is one of several redwood parks within easy reach south of the city. **Castle Rock State Park** is another destination known for its unusual rock formations and canyons. It is located on Skyline Boulevard, just south of Saratoga Gap where Highways 9 and 35 meet. The park's trails go through chaparral and oak woodlands, with distant coastal views. This combination of plant communities is also found in several large parks south and west of San Jose, including **New Almaden Quicksilver County Park** and **Henry W. Coe State Park.**

THE NORTH COAST

The North Coast is a thinly settled part of the state with national and state parks that include coast redwood forests, oak woodlands, chaparral, and coastal vegetation. The usual approach is north from San Francisco across the Golden Gate Bridge, on Highway 101. Driving straight through to Crescent City, the location of the headquarters of the Redwood National Park, takes about six hours, but an overnight stop allows a more relaxed pace. The road consists almost entirely of freeway and first goes through attractive hilly countryside with pretty suburbs, oak woodland, and chaparral. After about two and a half hours, mixed hardwood forest and coastal redwoods predominate. About six miles (10 km) north of Garberville, it is worthwhile leaving the freeway for the quieter parallel route called Avenue of the Giants, which runs through coastal redwood forest along the South Fork of the Eel River. This route traverses **Humboldt Redwoods State Park**, the largest of the state redwood parks. Its visitors center is close to the self-guided trail at the Founders Grove in the Rockefeller Forest, reached from the turnoff about 20 miles (32 km) after the beginning of Avenue of the Giants. Many other hikes are available in the park. After another 11 miles (17 km), the Avenue of the Giants rejoins the freeway, and within a drive of about one hour there are a number of overnight options, particularly in the towns of Fortuna, Eureka, and Arcata.

On the next morning, it is an easy drive to the Redwood Information Center (phone:

707-488-3461) of the **Redwood National Park**, near the town of Orick, where there are exhibits, publications about the park, and trail maps. There is good beach walking and tidepooling nearby. A small distance inland is the Lady Bird Johnson Grove, where there is a short interpretive trail. There is also a trailhead to the 8.5 mile (14 km) long Tall Trees Trail, which leads to a creekside grove of some of the world's tallest trees. The 13 mile (21 km) long Bald Hill Road also takes one through coastal redwood forest, followed by groves of Oregon, canyon, and black oak as it rises to Schoolhouse Peak.

Most of the 40 or so miles (65 km) along the coast between Orick and Crescent City to the north is part of the Redwood National Park. This part of the park has many scenic coastal trails and drives. In Crescent City, the main Park Headquarters at 2nd and K Streets has exhibits, information, maps, and publications. The Howland Hill Road from Crescent City to Stout Grove is a scenic, narrow gravel road that is lined by enormous trees. The return is by the paved main Highway 199.

For the return to San Francisco, one option is to drive 175 miles (290 km) south of Crescent City on the mostly freeway Route 101 to Leggett, and there turn off on the slower and scenic winding coastal Route 1. The area between between Fort Bragg and Albion has particularly picturesque coastal scenery, six state parks, and many attractive bed-and-breakfast inns. The easiest return to San Francisco is via Route 128, which joins Highway 101 at Cloverdale.

THE SIERRA FOOTHILLS

The Sierra foothills east of the Central Valley have a number of small towns that were founded after the gold rush of 1849 and that comprise the Mother Lode or Gold Country. Many Californians drive through this part of their state en route to the more popular attractions of Lake Tahoe and Yosemite National Park without stopping for more than fuel and a meal. Side roads of the Mother Lode give a more relaxed atmosphere, and its towns make a good base for visits to nearby Calaveras Big Trees State Park, Merced Grove, and Yosemite Valley, which should not be missed if one has never been there. This part of the state is good for a visit almost any time of the year, though less so in the summer when it becomes rather baked. In late winter and spring it is at its best lush green with full streams and abundant wildflowers against a background of the still snowy Sierra. In autumn, there is mild weather in which to enjoy the fall color of the black oak, big leaf maple, and other deciduous trees.

The **Mother Lode** towns, which retain many of their 19th-century buildings, are linked by the appropriately numbered Highway 49. This is one of the state's best known oak viewing highways and also includes areas of chaparral. The part of the highway between Placerville and Sonora is probably of greatest interest for its pleasant towns, which include Sutter Creek, Jackson, Murphys, and Columbia, which is the best-preserved gold mining town in California and a State Historic Park (Fig. 10.02). The roads of the Mother Lode take one through a pleasant countryside with canyon and interior live oak lining the gorges, hillsides dotted with blue oak, and ridges topped with black oak and pine.

FIG. 10.02. *Livery stable in Columbia State Historic Park, the best-preserved gold mining town in California.*

Calaveras Big Tree State Park is a short distance from Highway 49 at Angels Camp. One turns east on Route 4 and continues about three miles (6 km) past the town of Arnold. Although the park is best known for its truly majestic giant sequoia trees or sierra redwood (*Sequoiadendron giganteum*), it also has a trail leading through groves of canyon and black oak. The Calaveras South Grove has the largest of the groves of giant sequoia in the park. These close relatives of the coast redwood are the world's most massive trees. The grove is in a completely natural state and is regarded as one of North America's most magnificent forests (Dewitt, 1993). It is reached by a two-mile trail from the Beaver Creek picnic area.

One of the most attractive approaches to **Yosemite National Park** is from Mariposa, the southern terminus of Highway 49, on Route 140, which follows the deep canyon of the Merced River for most of the way. In the canyon are groves of canyon and interior live oak growing among a variety of conifers. Yosemite Valley itself has groves of black oak along meadows in a dramatic setting of granite cliffs and waterfalls. The walk between the main visitors center and the base of Yosemite Falls also has many oaks. There is abundant flow of water in the falls and river during the spring snow melt, which typically peaks in May. Leaving the valley by Route 41 provides some spectacular views from a turnout after one has gained some altitude. About 24 miles from the valley is the southern entrance to the park and a turnoff to the impressive **Mariposa Grove** of giant sequoias.

After leaving Yosemite National Park, Route 41 traverses first a landscape of canyon and live oak mixed with conifers and then descends into rolling blue oak woodlands and savannas before reaching Fresno and the Route 99 Central Valley freeway. The Route 198 turnoff at Visalia lies less than an hour's drive south on Route 99. Visalia has made a special effort to preserve some of the groves of valley oaks that once surrounded the town. The nearby **Mooney Grove Park** was established in 1909 and preserves many huge old valley oaks. Route 198 leads east from Visalia to the **Sequoia and Kings Canyon National Parks**. In Sequoia National Park, the middle fork canyon of the Kaweah River has fine oak woodland before one reaches the groves of giant sequoias for which the park was established. During most of the year, when the road is not closed by snow, one can return to Fresno via the Generals Highway, which climbs into mixed conifer forest on its way to the Grant Grove of giant sequoias in Kings Canyon National Park. Route 180 westbound then takes one back to Fresno.

THE CENTRAL COAST

This region includes the popular Monterey-Carmel area and the Big Sur coast and its wild hinterland. Along with the North Coast, the Central Coast is an area in which one can see all of the major plant communities described in Chapter 5. However, the star attraction on the Central Coast is its varied coastal vegetation and scenery.

About five miles south of Carmel, off Highway 1, **Point Lobos State Park** is regarded as one of the most spectacular pieces of coastal scenery in the world. There are rocky cliffs and coves, small beaches, and a picturesque and varied combination of coastal vegetation (Fig. 10.03). This is the only area where Monterey pine and Monterey cypress grow together naturally. Because of the popularity of Point Lobos State Park, the number of visitors is restricted, making early mornings the best time for a visit.

Carmel lies at the beginning of the rugged and spectacular section of central California coast known as the **Big Sur** and traversed by Highway 1. The road is one of California's most popular tourist destinations, especially in the summer. In other seasons, nature travellers can have relative solitude on many of its varied trails. Much of the Big Sur is protected in the **Los Padres National Forest** and its **Ventana Wilderness**, where there are many of the longer trails. There are also a number of smaller state parks along the coast, where there are many shorter trails not only through coastal scenery along cliffs and to beaches, but also

FIG. 10.03. Rugged coastal scenery at Point Lobos State Park.

shady canyon walks into groves of coast redwood and oak. For anyone spending a few days in this area, the book, *Natural History of Big Sur* (Henson and Usner, 1993) is a good investment. In addition to providing a wealth of information on the flora, fauna, geology, and history of the region, there are almost 100 pages of trail descriptions that include the wildflowers, trees, and shrubs.

Garrapata State Park starts only about six miles (10 km) south of Carmel on Highway 1. It is poorly marked, but includes 4 miles of coast with highway turnouts near numbered gates. After the highway crosses the mouth of Soberanes Canyon, there are marked trails into the coastal vegetation of the headlands from gates 13, 15, and 16. Another trail leads east into the canyon, opposite gate 13, into redwood forest with mosses and ferns. One can continue on this trail to climb an open ridge through coastal scrub and chaparral for an elevation gain of about 2,000 feet (600 m) with views of the coast below. With a different return route, this makes a moderately strenuous 6 mile (10 km) loop.

Andrew Molera State Park, the largest of the Big Sur state parks, is a little further south on Highway 1, about 22 miles (35 km) south of Carmel. The long stretch of Molera Beach provides one of the most beautiful beach strolls along this coast. The cliffs behind the beach are rich in wildflowers for much of the spring and summer. Above the beach is an easy, almost level walk of about 2 miles (3 km) along the bluffs, through coastal scrub and grassy areas. Other trails flank the Big Sur River and enter a redwood forest community that extends into Pfeiffer Big Sur State Park.

Pfeiffer Big Sur State Park is about 26 miles (43 km) south of Carmel. This is another popular park with numerous trails. The Oak Grove trail is named for a coast live oak forest, but also traverses redwood groves and chaparral in a walk that takes a little over an hour. A few miles to the south is the **Julia Pfeiffer-Burns State Park**, which features a much photographed waterfall that plunges directly onto the beach at the mouth of McWay Canyon. The short Waterfall Trail leads through coastal scrub and coast redwoods to an overlook. Another short trail leads into the coast redwoods up the canyon.

Inland from the Big Sur coast is the **Hunter-Liggett Military Reservation**, an area noted for its oak woodlands, chaparral, and wildflowers. It is important to phone ahead to make sure that the roads are open (408 385-2403 or 408 385-5434) because the gates are closed for maneuvers for a few hours several times a month. The fastest approach is by Freeway 101, about an hour south of Salinas (near Monterey and Carmel) to the Jolon Road exit near King City. Jolon Road (G-14) leads to Jolon, where one picks up Mission Road to the Mission San Antonio de Padua. This large, well-restored mission was founded in 1771. Fine examples of valley oak savannas are found after turning off Mission Road onto Del Venturi Road and then continuing left after reaching Milpitas Road. This takes one to the northern border of the reservation, just beyond which are groves of blue, coast live, and valley oaks. After returning to Mission Road, a loop can be made to Highway 1 on the coast by taking the Nacimiento Fergusson Road about three miles (5 km) north of Jolon. This is the only road that crosses the Santa Lucia Coast Range of mountains. There are many switchbacks and several turnouts providing broad views of the coast. After reaching Highway 1, one can either turn north, returning to Carmel, or south toward San Simeon, Morro Bay, and Santa Barbara.

SOUTHERN CALIFORNIA

Southern California has several botanical gardens that are devoted to native plants. One of these is the **Santa Barbara Botanical Garden,** which features areas of oak woodland and chaparral communities. Docent tours are offered and there is an excellent bookstore and library. A pleasant drive from Santa Barbara is to the top of **Figueroa Mountain** via Highway 154 to Los Olivos, turning right on Figueroa Mountain Road. This takes one first through valley oak savanna, and then uphill into blue oak woodland, followed by scrub oak chaparral. The final two miles (3 km) on the unpaved Figueroa Lookout Road reach the 4,500-foot (1,350 m) summit. The wooded peak has excellent views of the surrounding countryside and is known for its wildflowers.

In the **Los Angeles-Orange County** area, the **Rancho Santa Ana Botanic Garden** is devoted entirely to California native flora. It is in the town of Claremont, off the San Bernardino Freeway (Highway 10). There is an oak identification display, and the Woodland Trail includes coast live oak and Engelmann oak. In Southern California, the coast is more densely settled than in other parts of the state, and there are relatively few coastal parks. One of the few in this highly urbanized region is **Crystal Cove State Park**, between the popular resort towns of Newport Beach and Laguna Beach. This park includes three miles of beach and a hinterland of scrub oak chaparral and coast live oak woodland. About a 30-minute drive inland is the **Ronald W. Caspers Wilderness Park**, one of the few wilderness areas in Orange County, with a choice of many trails through oak woodland and chaparral.

Further south, in the **San Diego** area, **Torrey Pines State Beach** is notable for its groves of Torrey pines (*Pinus torreyana*), probably the rarest of California pines. There is only one other small grove extant, and it is found on Santa Rosa Island off the coast near Santa Barbara. Torrey Pines State Park consists of a coastal mesa which is cut by steep canyons that plunge down to the beach. From the mesa, one can look down on the asymmetrical, wind-shaped, and twisted Torrey pines below. The tree has enormous, heavy cones with sharp, hooked scales.

The large **Cuyamaca Rancho State Park**, inland from San Diego, is reached by taking the I-8 freeway for 30 miles to the east, and then driving north for 12 miles on Highway 79 to the park headquarters. Another three miles (5 km) to north, there is an interpretive center. The park has the largest number of species of oaks and conifers in Southern California (Pavlik et al., 1991). The visitors' center provides trail maps and has an Indian Museum. Stones with holes used by the Indians for grinding acorns are found in many parts of the park. There is a variety of easy and moderate trails through chaparral, woodland, forest, and wildflower meadows.

Further north, about halfway between Los Angeles and San Diego is the **Santa Rosa Plateau Ecological Reserve**, established primarily for preserving an outstanding savanna of Engelmann oaks (*Quercus engelmannii*), which has an array of meadows and vernal pools with glorious spring wildflower displays. The reserve is near the Clinton Keith Road exit of the I-15 freeway between Lake Elsinore and Murrieta.

B. CHILE

Almost all visitors enter Chile by air and arrive in Santiago, a modern, bustling city with more than four million people in its metropolitan area. It lies at the center of the mediterranean climate area of central Chile. The best season for wildflowers is early spring, September and October, when it may still be possible to include skiing in the nearby Andes as an added attraction. A good season for travelling south to the araucaria forests and the scenically stunning Lake District is in November and December, when rain is diminishing. But the sunniest weather is from January through March, which are also the summer vacation months when accommodations are most crowded.

The national parks and nature reserves of central Chile are the best places to see the vegetation of its mediterranean climate region. They are not heavily visited, trails are generally unmarked, and some of the roads require four-wheel drive vehicles. Most of the parks are not served by public transportation, and tourist agencies do not generally feature these destinations. Even with a rental car, the paucity of signs on gravel and dirt roads requires periodically stopping for directions. These circumstances make some of the tours organized by North American and British botanical gardens and specialists in nature travel an attractive option for Chile.

The most recent edition of the Lonely Planet Guide for Chile is good in providing more information than previously on national parks. Even better is the excellent guide book entitled *Chile, A Remote Corner on Earth, Tourist Guide*, published in 1992 for the World's Fair in Seville by Turismo y Comunicaciones S.A. In Chile, this was still available late in 1995 and is well worth taking the trouble to find for its excellent descriptions and detailed maps. It was based on a series revised annually and published in Spanish by the telephone company. These Turistel Guides include one for the Santiago region (Central), another for the north (Norte), and a third for the south (Sur). There is also a combined edition and a camping guide. Another part of the Turistel series is a map in booklet form, with each two-page spread showing one of the north-south segments of the country. This format is perfectly suited to the long, narrow shape of the country, and avoids the awkwardness of unfolding maps in the car. The Turistel guides and maps are available in most large Santiago bookstores and at many newsstands. Also worth mentioning is the *South American Handbook* (Trade and Travel Publications in Britain and Prentice Hall in the United States). It is also revised annually, and has a wealth of information on accommodations in all price categories, restaurants, bus schedules, etc., but there is little coverage of the national parks and nature reserves of central Chile.

For information on plants, the excellent field guides by Adriana J. Hoffmann come in a convenient format with superb color illustrations. Although the text is in Spanish, the books are clearly organized with descriptions and botanical terms that are often similar to the corresponding English words. The most useful of the guides for the mediterranean climate part of Chile is *Flora Silvestre de Chile, Zona Central,* which is devoted primarily to flowers. A companion volume, *Flora Silvestre de Chile, Zona Araucana,* concentrates on trees, shrubs, and vines of the rainier, southern parts of central Chile. These books are also widely available in large bookstores. They are also carried by the Gaia Bookstore, which is devoted entirely to books on ecology and is located on Orrego Luco, a half block north of its intersection with Providencia. Gaia carries a series of very compact pocket guides that is bilingual in Spanish and English in the Coleccion Naturaleza de Chile (Chilean Nature Collection). The booklets entitled *Arboles Nativos de Chile* (Chilean Trees) and *Arbustos Nativos de Chile* (Chilean Shrubs) are by Claudio Donoso. Another, *Plantas Trepadoras del Bosque Chileno* (Chilean Vines) is by Martínez.

In Santiago, general tourist information can be obtained from Sernatour, the national tourist office (Av. Providencia 1550; Tel: 698-2151) and from private travel agencies. The Automóvil Club de Chile (Vitacura 8620; Tel: 212-5708) extends benefits to foreign auto-mobile club members. In addition to having maps and guides, it has a car rental agency with branches throughout the country. It may be worth stopping at the CONAF (Corpo-ración Nacional Forestal) office (Presidente Bulnes 259, office #206; Tel: 696-0783) to get small brochures about individual National Parks that may not available at the parks them-selves. The parks are not as heavily visited in Chile as they are in California and Europe. Since spring is not a major travel season, it is worth making sure that a park is open when you hope to visit. Some may be staffed only on weekends.

The Chilean airlines, Lan-Chile and Ladeco, offer substantial discounts on tickets for domestic travel that are bought outside the country. These are well worth looking into, given the fact that rental cars are available in many cities such as La Serena to the north and Temuco, Valdivia, and Puerto Montt to the south. Santiago, with its heavy traffic and many one-way streets, may be difficult for the visitor, but driving conditions are easier on paved roads outside the city. Access to national parks is generally by gravel roads, which are sometimes in poor condition, especially after rain, and in some cases only suitable for high chassis, four-wheel-drive vehicles. A bit of Spanish is a big help in travelling indepen-dently. Chile is full of inexpensive bed and breakfast inns. They are often cozy and provide a better way to get to know the country than the bigger hotels.

Santiago Area. Less than one hour south, just beyond the outskirts of Santiago, is the **Río Clarillo National Reserve**. It occupies the scenic canyon of a branch of the Maipo River in the Andean foothills and includes sclerophyll forest and matorral. There is a new visitors center and a small arboretum with labeled examples of many of the shrubs and trees of central Chile. Nearby is a picnic area by a clear stream that is good for a refreshing dip. Within a short drive of the park is the main valley of the rushing Maipu River, which penetrates higher into the Andes with views of snow-covered peaks. There are several inns offering attractive accommodations, some with large swimming pools and river and moun-tain views.

La Campana National Park is the outstanding destination for seeing almost all of the plant communties of the mediterranean climate part of Chile in a relatively undisturbed form (Rundel and Weisser, 1975). It has northern and southern entrances, which are not connected by road. The park comprises an area of about 20,000 acres (8,000 hectares) including some of the highest peaks in the Coast Ranges. The northern entrance leads to what are probably the largest and most beautiful remaining groves of Chilean palms, surrounded by matorral and sclerophyll woodland. This area can be approached by taking Route 5 (the paved Pan-American or longitudinal highway) north from Santiago and turning off at the village of Ocoa. The southern approach to the park can be reached by continuing on route 5 to the Quillota turnoff. An alternative route from Santiago is route 68 to the coast through Viña del Mar to the pleasant fruit-growing and holiday town of Olmué. The town is close to the south gate of the park and has comfortable accommodations. From the south gate, a challenging, narrow dirt road winds uphill through sclerophyll forest and matorral, followed by *Nothofagus obliqua* forest at higher elevations. La Campana is about a two-hour drive from Santiago.

To the North. About three hours north of Santiago on Route 5 is the Catapilco turnoff, which leads to the three coastal resorts of **Zapallar**, Cachagua, and Papudo. The Carmel-like atmosphere of Zapallar provides a convenient base, with a pleasant waterfront promenade, beaches, and outdoor seafood restaurants. On this section of the Pacific, the Coast Range drops steeply to the sea and retains much of its original vegetation of sclerophyll forest and matorral. The **Aguas Claras Ravine** is a nature reserve that lies two miles (3 km) south of Cachagua on an unmarked turnoff. A dirt road extends up into the hills after a shady, forested drive near a brook.

La Serena lies near the northern limits of the mediterranean climate region. This charming old town of about 100,000 people is about 300 miles (500 km) north of Santiago. La Serena is a popular vacation destination for Chileans, many of whom come in the summer to enjoy the nearby beaches. Accommodations are not a problem in spring, which is off-season. There are good air and bus connections from Santiago and there are car rental agencies at the airport and in town. **Fray Jorge National Park** with its relict cloud forest provides a spectacular coastal setting, and is less than a two-hour drive from La Serena. The park has been declared a UNESCO International Biosphere Reserve because of its unusual setting of a small cloud forest surrounded by semi-arid vegetation. There are camping facilities in the park or one can stay at the pleasant hot spring resort of Termas de Socos, which is nearby. Also close to Fray Jorge National Park is the **Valle del Encanto**, a stream valley with interesting rock formations, petroglyphs, and grinding stones from the pre-Columbian culture (Fig. 10.04). A return to La Serena via Ovalle and Vicuña makes

FIG. 10.04. *Grinding stones used by pre-Columbian peoples to grind grain at Valle del Encanto.*

FIG. 10.05. *Valle de Elqui near La Serena.*

a pleasant circuit, which includes views of the rich, irrigated agricultural strip along the Limarí River. Many visitors come to La Serena to drive up into the beautiful Andean Valle de Elqui (Fig. 10.05), a scenic agricultural valley that at higher elevations is reminiscent of the Himalaya.

South to Temuco. This part of Chile is at the wet end of the mediterranean climate zone. It contains enormous plantations of Monterey pine (*Pinus radiata*) and also some interesting national parks that preserve native forests of deciduous and evergreen southern beech (*Nothofagus* spp.), and monkey puzzle tree (*Araucaria araucana*). Temuco is a city with a population of about 225,000 that is 400 miles (675 km) south of Santiago. Its good air and bus connections and car rental facilities are worth considering because the driving time from Santiago is about ten hours on a road with heavy truck traffic. **Nahuelbuta National Park** lies in the Coast Range, about a three-hour drive to the northwest of Temuco. The park includes a 4,750-foot (1,450 m) high peak. There are forests of the evergreen southern beech (*Nothofagus dombeyii*), deciduous southern beech, and the prehistoric-looking forests of *Araucaria araucana*. There is an interpretive center and a system of roads and trails.

In the Andes, there is the additional attraction of the ski resort of Llaima at the foot of the 10,000-foot (3,050 m) high volcano of the same name. Llaima lies between the national parks of **Los Paraguas** and **Conguillío**, both of which have *Araucaria* forests and rugged mountain scenery. It is worth making inquiries at CONAF at the corner of Caupolicán and Bulnes in Temuco about local road conditions and whether campsites are open. Spring rains are common in this area, and renting a four-wheel-drive vehicle may be advisable.

Valdivia and Puerto Montt. These two cities, each with somewhat over 100,000 people, are in attractive settings and are easily reached by air or bus from Santiago. An alternative is to go by train, in beautifully maintained Pullman sleeping and dining cars from the 1920s and 1930s, but over a somewhat bumpy roadbed. Puerto Montt is the more distant of the two cities at 650 (1,080 km) road miles from Santiago. Both are good starting points for visiting the national parks of the Lake District, known for snow-covered volcanos, deep blue lakes, and dense forests. This area may be rainy in the spring and even in the summer, but its spectacular scenic attractions make it hard to pass up if one has come far enough south to see the *Araucaria* forests. Though some rainfall is likely, it becomes less frequent toward December. Dense temperate rainforest can be easily reached by paved road to Puyehue and to Vicente Pérez Rosales National Parks. These are among the most popular national parks in Chile, but they are not heavily visited outside of the January to March summer season. One of the newer parks is Alerce Andino National Park, only 25 miles (40 km) east of Puerto Montt. Several tourist agencies in Puerto Montt arrange day trips to the park, which protects areas of alerce forest (*Fitzroya cupressoides*).

C. WESTERN CAPE, SOUTH AFRICA

The most recent Lonely Planet guide to *South Africa, Lesotho & Swaziland* includes generous coverage of parks and nature preserves and is a good introduction to the Western Cape. There is also an excellent series of compact, paperback *South African Wild Flower Guides*, which has four volumes devoted to the Western Cape. The area near Cape Town is covered in *Cape Peninsula* by Kidd (volume 3). Volume 5, *Hottentots Holland to Hermanus* by Burman and Bean is perhaps the most useful of these volumes in that it covers the fynbos area that has the greatest species diversity. It also briefly describes the best known hiking trails in this mountainous region. Volume 1, *Namaqualand & Clanwilliam* by Le-Roux, covers the famous semi-arid wildflower region north of Cape Town, partly beyond the mediterranean climate area proper. Volume 2, *Outeniqua, Tsitsikamma & Eastern Little Karoo* by Moriarty, includes the eastern part of the fynbos region, where it gradually enters an area of year-round rain. This part of the Cape Region includes forests on the ocean-facing, southern slopes of the mountains and the semi-arid Little Karoo on the inland, northern side. All four of the wildflower guides are available in the well-stocked bookstore of the spectacular Kirstenbosch Botanical Garden in Cape Town. Alternatively, they can be shipped at a somewhat higher cost by Honingklip Book Sales, 13 Lady Anne Avenue, 7700 Newlands, Cape, South Africa, which specializes in overseas sales of botanical and horticultural books.

The South African Tourism Board (Satour) is an excellent source for free literature. Their offices can be reached by 800 phone numbers: (800) 822-5368 for New York and (800) 782-9772 for Los Angeles. Among the most helpful free booklets for planning a wildflower trip are ones entitled *Western Cape* and *Northern Cape and Namaqualand*. Satour also can provide a one-page information sheet from Flowerline, a non-profit wildflower organization that provides "up-to-the-minute" information on what is in bloom and where, once one is in the Western Cape. Another useful brochure is *Follow the Footprints: South Africa's Hiking Trails*. Many day hikes are listed for the Western and Southern Cape, with addresses and phone numbers for making inquiries and information about obtaining permits where required. There are even more trails listed that require one or more overnights, often taking advantage of well-equipped huts. If you are considering a side trip to a game park, ask for the *Wildlife and Safari Country* booklet or contact one of the many travel agencies specializing in "safaris" to other nearby African countries.

The best months for a wildflower trip are August and September. Cape Town is ideally situated for exploring the small mediterranean climate area of the Western Cape. The city and its surroundings have a population of about two million. The spectacular scenery in and around Cape Town makes it as much of a destination in its own right as San Francisco is in California. The most frequent air connections are from Europe, usually via Johannesburg. Travel from the United States has recently been made much easier by non-stop flights to Cape Town from the eastern United States. Non-stop service to Johannesburg is convenient if your travel plans include Victoria Falls, Botswana and Namibia, and/or the closer and also famous game parks of the East Transvaal. South African Airlines has reduced fare tickets that make it possible to combine all three destinations with Cape Town at a relatively low cost within Africa.

Cape Town offers a large number of options for accommodations. Being near the center of the city is convenient if you are dependent on public transportation. If you rent a car, it is possible to stay at bed-and-breakfast inns in the nearby beach communities or the suburbs of Newlands and Constantia near the Kirstenbosch Botanical Garden. Car rental has the advantage of flexibility and easy day trips, but there are hotels in various price categories in the center of the city from which it is easy to book guided bus or van tours. Some of the agencies offering flower tours are Eco Explorers (021) 92-9361 or Fax (021) 930-5166, Springbok Atlas (021) 448-6545, Tailormade Tours (021) 72-9800, or Which Way Adventures (024) 852-2364 or Fax (024) 852-1584.

For the independent traveller, Cape Town is the place to pick up additional travel information while becoming acquainted with the city. Almost all of this is available in the center, within a few blocks of the main railroad station. Satour is in the Golden Acre shopping complex across the street from the station. Captour, which provides information only for the Western Cape, is in the nearby Strand Concourse. The National Parks Board is at the corner of Long and Hout and is the place for making reservations and getting information on national parks, including the Kruger National Park for game viewing.

Flowerline provides current information about the best places to see wildflowers in bloom. The office is located in the Tourist Rendezvous Travel Centre, Addersley Street, and the phone number is (021) 418 3705. It is worth asking for a schedule of the wildflower shows that are held both in Cape Town and in many smaller communities from late August to early October. They are a long-standing tradition that provides an opportunity to see many unusual species and to meet other wildflower enthusiast (Lighton, 1973). The Calendon wildflower show has been held since 1892, and the one in Darling is also venerable. The enthusiasm seen at these shows helps to explain the long-standing support for preserving native plants in the Western Cape and the existence of numerous botanical gardens and nature preserves.

National Hiking Ways, for hiking information, is on Martin Hammerschlag Way near the intersection with Oswald Pirow. Lastly, the Cape Nature Conservation office, which is responsible for the provincial parks of the Western Cape, is in the suburb of Bellville, and can be phoned for information (021) 948-7490.

Cape Town. The most familiar natural feature of Cape Town is the roughly 3,500-foot high (1,000 m) Table Mountain, which serves as a dramatic backdrop for the city and its suburbs. There is a cable car to the top of the mountain, where there are excellent walks through areas of rich fynbos flora. Mornings and early evenings are best for visibility, but it is worth calling (021) 24-5148 to check on weather conditions because dense fog and high winds are common. Being prepared with warm clothing and rain gear is a good idea in any case.

Another great attraction is **Kirstenbosch Botanical Garden**, considered one of the most beautiful botanical gardens of the world. It is of special significance because it was the first to cultivate and display native plants exclusively when it was founded in 1913 (Rycroft and Ryan, 1980). It is about 7 miles (13 km) from the center of Cape Town, close to the suburbs of Newlands and Constantia. A 90-acre (36 hectare) landscaped area on the lower, eastern slopes of Table Mountain is contiguous with an additional 1,200 acres (490 hectares) of natural fynbos, woodland, and forest that extend to the top of the mountain. Kirstenbosch is an excellent place to become acquainted with the plants of the Cape's mediterranean climate area as a whole before seeing them in their natural habitats. There is also a well-stocked bookstore and a pleasant restaurant that serves breakfast and lunch.

The Cape Peninsula includes the **Cape of Good Hope Nature Reserve**, which is an easy day trip from Cape Town. Starting your route clockwise on the eastern side of the peninsula makes it easier to stop at the side of the road for views (you drive on the left side of the road). In addition to abundant fynbos vegetation, there is varied wildlife. One is warned not to feed the baboons, which are an occasional roadside attraction, but are unpredictable and dangerous. The return to Cape Town is via the famous cliffside coastal route of Chapman's Peak Drive, via the spectacular harbor of Hout Bay.

The West Coast. North of Cape Town, some of the best places to see the strandveld vegetation are within a one- to two-hour drive. The small town of Darling has three floral reserves in its vicinity. The large **West Coast National Park** is known for its flowering annuals and succulents in its Postberg section. It is the largest area of protected strandveld, but much of it is closed to hikers. Slightly north of the National Park, near the town of Paternoster is the **Columbine Nature Reserve**, named for Cape Columbine and visited for its flowers and coastal scenery. One can visit the strandveld on a day trip from Cape Town or stay overnight in the coastal towns of Langebaan and Paternoster. Alternatively, one can drive inland to stay in Clanwilliam, a town at the foot of the **Cedarberg Mountains**, which has nearby nature reserves and a large wilderness area. From there one can continue north to visit the spectacular floral displays in semi-arid Namaqualand.

Namaqualand. This area of the Western Cape is beyond the dry end of the mediterranean climate area. It is known for its vast fields carpeted with daisies (Asteraceae) and vygies (Mesembryanthemaceae) after a good winter rainy season. The best time to visit varies with the timing of rain, but is usually in late July, August, or early September. It is best to view these flowers during the middle of the day, when they are open, and with your back to the sun so that the flowers are facing toward you. The towns of Vanrhynsdorp and Calvinia in the southern part of Namaqualand and Kamieskroon and Springbok further north can serve as places from which to explore this region of succulents and annuals. All of these towns can be reached by bus, but this area is best explored by car. Kamieskroon is about 300 miles (500 km) north of Cape Town on the excellent paved route N7. Nearby is the spectacular **Skilpad Wildflower Reserve**, with carpets of daisies of various colors against a backdrop of blue-green mountains (Fig. 7.15A). The flowers are fully open only from about 10 AM to 3 PM, allowing time for a leasurely breakfast before getting underway.

Winelands. This region includes many of the oldest wine growing areas of the Cape. The Winelands are only 30 miles (50 km) by freeway from Cape Town, making a suitable destination for a day trip. The region is known for its old Cape Dutch towns, historic small inns, wineries, and good restaurants. The Winelands' beautiful mountain scenery makes the area a pleasant base for a stay of several nights. Stellenbosch is the oldest and one of the best preserved old towns, and the seat of an Afrikaans-language University. It has several charming small inns and is a good town for strolling. Less than a half hour drive away from Stellenbosch is the town of Paarl, known for the **Paarl Mountain Nature Reserve**, with its mountain fynbos vegetation, including many species of *Protea*. It has a cultivated wildflower garden and has numerous hiking trails with excellent views. The near-

FIG. 10.06. Winelands. View from the terrace of a restaurant toward the wine village of Franschhoek.

by village of Franschhoek is set at the end of a beautiful valley (Fig. 10.06), with a backdrop of mountains that are sometimes snow topped during winter. The large **Hottentots Holland Nature Reserve** can be approached from Stellenbosch or the scenic road to Franschoek Pass. It is described in the *Hottentots Holland to Hermanus* volume of the South African Wild Flower Guides. There are short hikes or a three-day route for backpackers through varied mountain scenery and fynbos vegetation. Permits for either have to be obtained in advance from the Branch of Forestry, Private Bag 9005, Cape Town.

The Overberg. This is a large scenic region that includes the most varied mountain and coastal fynbos vegetation (Fig. 10.07). Hermanus is a convenient and beautiful base for exploring this area. The coastal towns of Betty's Bay and Kleinmond are close and have nature reserves with short, pleasant walks that are described in *Hottentots Holland to Hermanus*. The **Harold Porter Botanic Gardens** in Betty's Bay stretch up the mountainside to a waterfall near which the rare disa orchid (*Disa uniflora*) can be seen. The **Kleinmond Coastal Nature Reserve** has an 11.5 mile (18.5 km) trail through a coastal fynbos area with a large variety of birds. In Hermanus itself, there is the **Walker Bay Nature Preserve** with seven miles (12 km) of wild, unspoiled beach, and the **Fernkloof Nature Reserve** with abundant plant and animal life and 24 miles (40 km) of hiking trails. Walker Bay is also known as a spring calving site for the southern right whale. A short distance inland

FIG. 10.07. *Rugged coastal scenery within a two-hour drive of Cape Town.*

from Hermanus, over Shaw's Pass to Calendon are the **Calendon Gardens**, the site of a noted Flower Festival in mid-September. Further east is the large **De Hoop Nature Reserve**, a 30-mile (50 km) drive on good gravel road from the towns of Bredasdorp or Swellendam.

Karoo. The relatively dry, interior area around the Breede River Valley can be reached in about two hours from Cape Town. In addition to several wildlife and nature preserves, it has the famous **Karoo Botanical Garden** near Worcester, specializing in rare succulents and including an area of natural, semi-desert vegetation.

The Garden Route extends eastward from Mossel Bay. This coastal region with white, sandy beaches has South Africa's only temperate rainforest on the south-facing slopes of the Outeniqua and Tsitsikamma Mountains. The centers of this district are the towns of George and Knysna. Here the rain is spread fairly evenly throughout the year, explaining the presence of lush vegetation. Only a short drive to the north over one of several spectacular mountain passes is the semi-arid Little Karoo and the town of Oudshoorn.

There are several round trips that can be made from the white sandy beaches and lagoons along the coast, through forest with their giant yellowwood trees and wildflowers of genera such as *Protea*, *Gladiolus*, *Erica*, and *Watsonia*. After crossing the Montague or Prince Alfred's Pass (both on gravel roads) driving north into the Little Karoo, one can return south by the paved national highway over the Outeniqua Pass. This area includes fynbos vegetation in close proximity to the succulents and annuals of the Little Karoo and the woodlands and forest toward the coast, protected by the large **Wilderness National Park**, **Goukamma Nature Reserve**, **Knysna National Lake Area** and the **Tsitsikamma Coastal National Park**.

D. Western Australia and South Australia

A good way to start planning a visit is to browse through the Lonely Planet guidebook for Australia, which is both readable and full of useful factual information about accommodations, restaurants, and public transportation. The Australia-based series was originally geared to the backpacker, but is now written for a much broader public. For those travelling mainly in Western Australia, the Lonely Planet guide, *Western Australia*, is the best choice for its more extensive information and focus on natural history. In this respect, it is one of the best in the series. The Australian Tourist Commission office in the United States (c/o OTI, 1000 East Business Center Drive, Mount Prospect IL 60056. Tel 708 296-4900) has good general purpose brochures about Australia.

1. Western Australia

The Western Australian Tourist Centres (Forrest Place, Perth, Western Australia 6000) has booklets on accommodations, camping, and attractions in various parts of the state. An outstanding one, entitled *Wildflower Discovery: A Guide for the Motorist*, is designed for those who plan to rent a car and travel independently. It provides a choice of several detailed itineraries, with suggestions on where to stop and what flowers are likely to be found. Some of the recommended tours include the drier, flower-rich area north of Perth, toward Geraldton and Kalbarri, whereas others are for the moister, southwestern portion of the state which includes *Eucalyptus* forests and woodlands, coastal scenery with kwongan, and mallee.

Spring wildflower tours are one of Western Australia's major tourist attractions, and there are many guided tours originating in Perth. The Western Australian Tourist Centres, for example, carries a brochure, *Westrail Wildflower Tours of Western Australia* , which lists a variety of 3 to 8-day bus tours that are escorted by a naturalist. The tours include meals and accommodations at an average cost of about $100 U.S. per day. Additional touring companies offering wildflower tours are listed in the Lonely Planet Guides.

Many Australians are sports and hiking enthusiasts. There are abundant campsites and inexpensive hotels for backpackers. In Western Australia, the Stirling Range National Park and Porungurup National Park near Albany are among the most popular destinations for hikes, and there are numerous wildflower day hikes near Perth. Hiking is called bushwalking. Bushwalking clubs include Bushwalker of Western Australia, (09) 387-6875 and Perth Bushwalkers (09) 362-1614, and can be called for advice after arrival. Brochures about trails and campgrounds in several of the national and state parks can be obtained from the Western Australia Department of Conservation & Land Management, 50 Hayman Road, Como, Western Australia 6152. In 1988, a well marked series of Heritage Trails was inaugurated for visiting historical as well as natural areas of interest. Information about these can be obtained from the Western Australia Heritage Committee at 184 St. George's Terrace, Perth 6000, Western Australia; Telephone (09) 322-4375.

From mid-August through October, the temperature is ideal, with daytime highs of about 65 to 70°F (19 to 21°C) around Perth, somewhat warmer to the north and cooler to the south. Rainfall peaks in July and then tapers off. Rain on at least a part of some days is a virtual certainty, with an average of 14 days for September in Perth and 19 days in Albany. Rainfall is scantier, less frequent, and more variable from year to year toward the north, where wet years will bring spectacular carpets of wildflowers, but dry years may be disappointing.

It is worth inquiring about the dates of the two-week school holiday that is usually scheduled between late September and early October, especially if you are travelling independently. The dates are staggered among states. During school holidays, there will be more traffic on roads that are otherwise relatively empty, and accommodations in resort areas are likely to be scarce. At other times, it is usually easy to find vacancies in spring, except in Kalbarri, a popular vacation spot during the cool season, and in the Wheatlands, where accommodations are scarce.

Perth is almost invariably where visitors arrive, often after an almost five-hour flight from Sydney, the busiest entry point for Australia. A first night in Perth allows a rest from a long trip. Perth is a large, prosperous, and attractive city overlooking the curving Swan River. It has a famous botanical garden in Kings Park that specializes in native plants, and this is a good place to become oriented since many of the plants that will be seen in the wild are labelled here. In addition to the botanical garden, there are paths through parts of Kings Park that have been left with their original vegetation and are rich in wildflowers. Getting around the center of the city and to Kings Park is made easy by the availability of a free bus service in the central area. Bikes can be rented at the western side of the main parking area and are useful for exploring the many bike tracks in the park. There is also an attractive restaurant and coffee shop.

Another easy-to-reach and interesting area in Perth is the attractive and lively cultural center near the central railroad station. This area includes the Museum of Western Australia, which provides a well displayed introduction to the natural history of the state. Conveniently adjacent is the Northbridge neighborhood with its many good restaurants and pubs.

A visit to some bookstores is worthwhile, since there are many good books on the diverse flora of Western Australia. Among the most useful is *Flowers and Plants of Western Australia* by Erickson et. al (1991) cited in the bibliography. This beautifully illustrated book is organized by region, includes clearly written, informative text, and will be useful on the trip.

Detailed road maps and an accommodations guide are available at very modest cost from the Royal Automobile Club of Western Australia (RAC), 228 St. George's Terrace, on presenting a membership card of an automobile club. There is also a well stocked bookstore with an excellent travel section. The RAC has a well informed staff that can help plan itineraries and make reservations. Since conditions for wildflowers vary from year to year, it is worth obtaining up-to-date information from the RAC or the Western Australian Tourist Centres on Forrest Place.

For a visitor with at least 10 days to two weeks time, a good plan is first to make a loop north of Perth, where the weather is warmer and spring comes earlier. This can then be followed by a southern loop that includes forests, some dramatic coastline, and the mountain ranges near Albany. Most towns have tourist information offices, typically with extremely helpful and well informed staff. Motels generally have a small refrigerator and coffee and tea makings in each room. Freezing a partially filled plastic bottle can provide cold drinking water on walks the next day. Many places also have a washer and dryer for the use of guests, making it possible to travel lightly.

North from Perth to Kalbarri. The northern loop is likely to take one at least as far as Kalbarri, a total of about 370 miles (620 km) in an almost eight-hour drive on the Brand Highway. With interesting stops along the way, this is best made as a two- to four-day trip with a stop in the oceanfront towns of Dongara or Geraldton. **Moore River National Park** is a less than two-hour drive from Perth and has low woodlands of *Banksia* over varied shrubs. It has been described as one of the world's outstanding wildflower spots. **Coomallo Creek** is about halfway from Perth to Geralton and is another of the state's best wildflower sites, with over 200 species of plants.

Kalbarri is a popular resort with beautiful beaches and coastal cliffs. It is an excellent base for the nearby **Kalbarri National Park**, which encompasses the long **Murchison River**

Gorges, with their spectacular geological formations, wildflowers, and wildlife. Two of the most impressive areas, known as the Loop and the Z-Bend, are 20 miles (35 km) by road from Kalbarri. There are short trails down into the gorge and longer ones elsewhere in the

park. The paths and roads are lined by many species of *Grevillea* and *Hakea* and abundant feather flowers (*Verticordia* spp.). An unpaved road south of Kalbarri can be taken to an area of oceanfront cliffs and outstanding views.

On the outskirts of Kalbarri is the privately owned **Kalflora Botanical Garden**, an area of bushland with paths and interpretive signs identifying many of the plants. Wildflower tours available from Kalbarri are likely to include a stop at Kalflora.

FIG. 10.08. *Toodyay, a pleasant town for a stopover.*

The return route to Perth can be via the Wheatlands, which have their own characteristic flora. Driving inland from Geraldton, one reaches the "Wildflower Way," which extends south from Mullewa to Morawa. One can then can cut over to the Midland Road, taking it from Three Springs to Moora. Wildflowers are often best seen on exploring the relatively wide roadsides. These inland roads traverse an agricultural area with fewer parks and nature reserves than along the coast. The area is also known as a corridor for migratory birds. Towns are small and far apart with few accommodations, and having advance reservation for the night can make for a more relaxing day. A longer drive reaches the Avon Valley, east of Perth. There are pleasant stopovers here, such as the town of Toodyay (Fig. 10.08).

South from Perth to Albany and Augusta. A southern loop offers an even greater variety of routes, national parks, reserves, and overnight stops in greener, moister, and more wooded surroundings. Just to the southeast of Perth is the hilly **Serpentine National Park** with an abundance of plants and wildlife on trails that include a river with waterfalls. The Albany Highway is the most direct road to the **Stirling Range National Park**, which is about 200 miles (320 km), a drive of almost five hours southeast from Perth. This range rises to over 3,300 feet (1,000 m) from the surrounding plain and includes the highest peaks in the southwestern part of the state. The nearby **Porongurup National Park** is known for its dramatic granite domes (Fig. 8.14A). There are well marked but steep trails, requiring hikers about two hours to reach their roughly 800-foot (250 m) high tops. Areas of granite are surrounded and topped by a variety of flowering plants, and there is a beautiful karri forest on the northern side of this compact range. The small and pleasant town of Mount Barker is a convenient overnight stop for these two parks.

Albany is a sizable coastal town and an excellent base for the surrounding region. The Stirling Range and the Porongurups are less than an hour away. Tour agencies in Albany offer wildflower tours to both places. It is also near some of the world's most spectacular coastal scenery. **Torndirrup National Park** is on a rocky peninsula that protects the harbor of Albany. In addition to dramatic coastal cliffs, coves, and rock formations, it has an abundance of sundews, orchids, *Banksia*, *Dryandra*, and *Grevillea*. To the west of Albany, there are additional places to see varied coastal vegetation at **West Cape Howe**. The small town of Denmark is another pleasant place to stay to see the beautiful **William Bay National Park** and the karri forest in the Valley of the Giants.

To the west of Denmark, the main road turns inland at Walpole and is bordered by jarrah forests. The lumber town of Pemberton is a pleasant stopover. The tall, streamside karri forest of **Warren National Park** is nearby, with colorful flowering vines in the understory. West of Pemberton, the road is again lined by jarrah forest, and after a few hours' drive, one reaches the coast near where the Indian Ocean meets the Southern Ocean at Cape Leeuwin, near the resort town of Augusta. The coast road between Augusta, Margaret River, and Busselton goes through varied scenery with long beaches, cliffs, sheltered bays, and windblown coastal vegetation. The Caves Road takes one through an area riddled with limestone caves, among which is the spectacular Jewel Cave. Between this region and Perth, there is a wide choice of overnight stops.

On the way back to Perth, between Busselton and Bunbury is the **Ludlow Tuart Forest** (*Eucalyptus gomphocephala*), with stately 300 to 400 year old trees of a species that grows nowhere else. Bunbury is also a place where dolphins have regularly come in to be fed since 1990. Between Bunbury and Mandurah is **Yalgorup National Park**, which includes a large number of brackish, mangrove- and paperbark-lined lakes and estuaries near the coast. It also has unusually abundant birds and mammals.

2. SOUTH AUSTRALIA

South Australia offers many natural attractions, but is rarely the destination for a tour focused primarily on wildflowers. Visitors are more likely to be drawn by the rich animal life of Kangaroo Island, the wild, semi-arid landscapes of the Flinders Range, and the wine country in the Adelaide Hills. The capital, **Adelaide**, is known for its large parks, stately public buildings, and attractive suburbs in the hills. As in all of the Australian state capitals, a large botanical garden, **Botanic Gardens of Adelaide, South Australia**, is a centrally located amenity. The garden has a section for native plants and a well stocked bookstore where one can prepare for trips to the surroundings. The garden, museums, and the cultural center are all located on the broad and gracious North Terrace. The South Australian Travel Center at the corner of North Terrace and King William Street (18 King William Street, Adelaide 5000, South Australia. Tel: 08-212-1505) is a good place to pick up information and booklets about the Adelaide Hills, Kangaroo Island, and the Flinders Range.

For getting around the city by bus, train, or streetcar, it is economical and convenient to buy an all-day transit ticket. There are numerous parks in the Adelaide Hills that can be reached by bus on weekdays. **Belair National Park** is a pretty 30-minute train ride into the suburbs of the Adelaide Hills. The visitors' center is a ten- to 15-minute walk from the station and provides maps of the walking trails in the park. A four-hour walk with time for a picnic lunch allows one to get to the eastern part of the park, where there is relatively undisturbed mallee, woodland, and forest with a wide variety of understory plants. **Wittunga Botanical Garden** is also on the same suburban train line and specializes in Australian and South African plants arranged by genus.

A rental car allows the freedom of visiting the wine districts and staying at one of the many bed-and-breakfasts in the nearby hills. It is also the best and often only way to get to nature reserves, such as **Scott Creek Conservation Park** for woodland, mallee, and scrub and **Aldinga Scrub Conservation Park** for coastal vegetation. Both are within 30 miles (50 km) of Adelaide. It is a good idea to take water and lunch.

The outstanding destination for seeing relatively undisturbed woodland and mallee vegetation and abundant wildlife is **Kangaroo Island**, which is a 30-minute flight southwest of Adelaide. It can also be reached in a half day by driving from Adelaide and taking a car ferry. The western part of this 90 by 32 mile (150 by 52 km) island is protected in **Flinders Chase National Park**, for which land was set aside in 1919. This is the largest national park in South Australia, with 180,000 acres (73,000 hectares) or almost 17 percent of the island's total area. Wildlife includes koalas, sea lions, fairy penguins, echidna, emu, kangaroo (Fig. 10.09), and wallaby, all of which allow one to get close enough for good photographs. Kangaroo Island was never colonized by rabbits or foxes and its population remains only about 3,500. This has made it possible for 40 percent of the island to retain its natural vegetation, of

FIG. 10.09. *Kangaroo at a late afternoon meal on Kangaroo Island.*

which almost two-thirds is within the Flinders Chase National Park and 15 Conservation Parks (Robinson, 1985). Since much of the mainland around Adelaide has been modified by agriculture and logging, the island has become a popular destination, with package tours from Adelaide, bed-and-breakfast accommodations, naturalist guides, and rental cars all available.

Another popular destination is **Flinders Ranges National Park**, 200 miles (320 km) north of Adelaide. This is a rugged area with interesting rock formations, peaks to 3,700 feet (1,100 m), and abundant spring wildflowers after a wet winter. It is at the dry end of the mediterranean climate area of South Australia. Tours for small groups for either Kangaroo Island or Flinders Ranges, using four-wheel-drive vehicles, can be arranged in Adelaide or through Australian Travel Headquarters in the United States (800-546-2155).

E. The Mediterranean Basin

Planning a natural history trip to the Mediterranean Basin first requires making a choice among a multitude of possibilities. The most successful trips will generally cover only a small part of one of the more than 15 countries in the region. On a single trip to any of the four other mediterranean climate areas, it is possible to sample a representative range of parks and nature reserves from one end of the region to the other. This is clearly impossible in the much larger and more diverse Mediterranean Basin.

One of the distinguishing characteristics of the Mediterranean Basin is the pervasive manner in which the many national cultures have modified the landscape over the centuries. Old villages, historic shrines, monasteries, and castles are set into landscapes of olive groves, citrus orchards, and vineyards like works of art. Accommodations can be found in small hotels and inns that are often in restored old buildings that are furnished with local antiques. Tavern meals and picnics introduce one to delicious regional specialties. Nearby cities offer museums and cafes. All this is available among populations that have been accustomed to welcoming tourists for many generations. These many appealing features help to make up for the relative paucity of wild areas in the Mediterranean Basin.

The countries that have the widest range of mediterranean vegetation and that are also particularly well set up for tourists include Spain, Italy, and Greece. This section provides examples of destinations for trips to one or more areas in each of these countries. Just as one cannot visit more than a limited area, it is not possible to outline more than a few of the many options for a trip to the Mediterranean Basin.

Portugal, southern France, and Israel also have nature reserves and are well developed for tourism. Morocco and Turkey have had an increasing number of tourists over the last few decades, but nature travel has been promoted only recently. Visits particularly to these two countries, but also to the others, are often easier to make as part of a group. Many such options are available from Great Britain and the United States. The advantage of these trips is that the details of planning are taken care of, and that the naturalist guides generally know the best and least visited spots.

Books. As with the other mediterranean climate areas, books are the best starting point for planning a trip. The difficult choice of a destination is made easier by the excellent *Mediterranean Wildlife* by Raine (1994). This useful book gives brief descriptions of over 150 places around the Mediterranean Basin where plants, birds, and other wildlife are of special interest. For most sites, the author briefly describes common and unusual native plant species first. There are helpful maps, travel directions, and suggestions on where to find convenient and pleasant accommodations.

Other nature-oriented travel books for the countries of the Mediterranean Basin are hard to find. Most general purpose travel books emphasize urban, cultural attractions to the virtual exclusion of parks and nature reserves. Even books for trekkers rarely give much space to lower elevation areas of mediterranean vegetation or to topics of botanical interest. One series that attempts to fill this gap is published by the Sierra Club with the titles, *Wild Italy*, *Wild France*, and *Wild Spain*. The volume on Italy (Jepson, 1994) is the most helpful.

There are a few English-language wildflower books that deal with the mediterranean climate parts of the Mediterranean Basin as a whole. Perhaps the most useful is the large and comprehensive *Mediterranean Wild Flowers* by Blamey and Grey-Wilson (1993). The over 2,000 illustrations include small black-and-white drawings, accompanying the text, that help to identify species. In addition there is a large section of color illustrations for which Marjorie Blamey was awarded a gold medal from the Royal Horticultural Society. *Wildflowers of the Mediterranean* by the Schönfelders (1990) is more compact. It includes almost 1,000 species, with the descriptions conveniently placed opposite fine color photographs. An older but reissued book is *Flowers of the Mediterranean* by Polunin and Huxley (1990). It gives a readable and informative overview of the flora of the entire region.

Depending on the destination you have selected, it may be worth considering two outstanding wildflower books that deal with specific regions, both including Oleg Polunin as an author and published by Oxford University Press. Polunin was a well-known botanist and plant photographer who died in 1985. His works include *Flowers of South-West Europe: A Field Guide* coauthored with Smythies, 1991, and *Flowers of Greece and the Balkans: A Field Guide* (1997). Both are in print in paperback editions and are superbly organized. An introductory section outlines the terrain, climate, and the characteristics and origin of the flora. Next is a chapter entitled *The Plant Hunting Regions*, which describes the best sites for wildflower walks, scenic drives, and towns that make an attractive and convenient base for excursions. For each region, there are descriptions and drawings of the prominent flowers and suggestions for the timing of a visit. It rapidly becomes apparent that most of the sites for the lower elevation, mediterranean climate areas are best visited from March to May, when the higher elevation, alpine areas are still under snow. Also excellent is *Flowers of Greece and the Aegean* by Huxley and Taylor, 1989. This well written book discusses the flora by region and is helpful both for planning trips and as a field guide. It has the added interest of describing ancient uses of plants. Ancient plant lore is the focus of *The Greek Plant World in Myth, Art, and Literature* by Bauman (Timber Press, Portland. 1993).

Guided Tours. Great Britain has long had a large public interested in horticulture, botany, and nature travel. As a result, there are numerous special organizations and agencies offering botanical and natural history trips, many of which have Mediterranean destinations. Individual tours generally are limited to groups of 10 to 12 and are accompanied by a naturalist guide. A two- to three-week tour generally costs $1,500 to $2,000 including air fare from Great Britain. Most tour organizations cater to the nature traveller, offering a relaxed pace, less crowded destinations, and a degree of flexibility. Some examples are as follows:

- *Botanical Study and Natural History Trips* (29 Nore Road, Portishead, Bristol BS20 9HN, England. Tel: 275-848629. Fax: 275-843143)
- *Field Studies Council* (FSC) *Overseas* (Montford Bridge, Shrewsbury SY4 1HW, England. Tel: 1743-850164. Fax: 1743-850178)
- *Greentours—Natural History Holidays* (The Pines Farm, Biddulph Park, Biddulph,

Stoke-on-Trent, North Staffordshire ST8 7SH, England. Tel/Fax: 01260-272837).
- *Gulliver's Natural History Holidays* (Oak Farm (NW) Stoke Hammond, Milton Keynes MK17 9DB. Tel. 01525-270100. Fax 01525-270777).
- *Naturetrek* (Chautara, Bighton, Nr. Alresford, Hampshire SO24 9RB, England. Tel: 962-733051. Fax 962-733368).
- *Norfolk and Suffolk Wildlife Holidays* (72 Cathedral Close, Norwich NR1 4DF. Tel. 01603-625540. Fax 01603-630593)
- *Wild Horizons* or *Wild Wings*, International House, Bank Road, Bristol, BS15 2LX (Tel. 0117-961-0874).
- *Wildlife Travel* (Green Acre, Wood Lane, Oundle, Peterborough, PE8 5TP, England. Tel. 01832-274892. Fax 01832-274568).

For up-to-date offerings, it is worth writing directly or checking recent advertisements in British horticultural, wildflower, and natural history journals such as *Natural World*, the magazine of the Royal Society for Nature Conservation, *The Garden: Journal of the Royal Horticultural Society*, and *Quarterly Bulletin of the Alpine Garden Society*.

There are tours offered from the United States by museums, botanical gardens, and agencies specializing in botanical trips, such as Geostar Travel (800 624-6633).

Independent Travel. There are many places in the Mediterranean Basin where independent travel with a rental car makes it possible to combine outdoor botanizing activities with visits to other nearby attractions. Below are three examples that require a little driving and that have attractive options for meals and accommodations.

SOUTHERN SPAIN

The region of Andalucia is in the southeastern part of the Iberian Peninsula. It is easily reached through a busy international airport at Malaga. Andalucia has an enormous variety of landscapes within a small area, from coastal marshland and hillside maquis and garrigue to snow covered mountains that still have skiing when their lower slopes are in peak bloom. The region also retains hill towns, gardens, mosques, and palaces from the almost 800-year period of Moorish rule that ended in the 15th century. The greatest road distance from one end of Andalucia to the other is only about 300 miles (500 km), allowing one to visit several of its attractions without spending too much time driving.

La Doñana, Spain's largest and most famous national park, is best known for its salt marshes and lagoons, which are wet, except during the long, hot summer. The park is about 60 miles (100 km) from Seville and is the home of flamingos, spoonbills, egrets, herons, storks, and geese. It is one of the major stopping places for migratory birds between Africa and Europe, conveniently close to the Strait of Gibraltar, the shortest water crossing of the Mediterranean Sea. On higher ground are areas of maquis and garrigue, stands of stone pine, and groves of cork oaks that serve as nesting sites.

Cars are not allowed in the park. Half-day four-wheel-drive trips are offered from the interpretive centers at La Rocina and El Acebuche, both on the northwest boundary of the park. Reservations and permits for visits must be obtained from the Doñana Biological Station, Paraguay 1, Seville, Spain. A convenient place to stay is the Parador Cristóbal Colón, one of Spain's excellent government-operated inns. It is close to the park, near the town of Mazagón, surrounded by pines and lavender on a clifftop site overlooking the Atlantic Ocean.

The cities of Seville and Córdoba are easily visited before or after La Doñana, as there are good road connections. Both have important historical monuments and charming old neighborhoods. Córdoba was one of the most important cultural centers under the Moors in medieval times. Later, during the Golden Age of Spain, Seville was the country's largest city, prospering particularly in the 16th century from its monopoly on overseas trade with the Spanish colonies in the Americas.

La Ruta de los Pueblos Blancos. This "route of the white towns" connects a group of striking, whitewashed towns that were built by the Moors in attractive mountain settings southeast of Seville. The larger communities of Ronda and Arcos de la Frontera both have pleasant accommodations that can serve as a base for excursions to the Serranía de Ronda. Polunin and Smythies (1991) describe this mountain range as one of the most interesting and richest botanical sites in Spain and list some of the flowers that are to be seen on several excursions out of Ronda and the village of Grazalema. Near the latter is one of the few remaining forests of the endemic Spanish silver fir, *Abies pinsapo*, which formerly covered much of this mountain range. White towns that are also particularly attractive are Zahara de los Membrillos, Setenil, Ubrique, and el Bosque, which has an inn. These villages are mainly clustered north and west of Ronda.

To the north of Córdoba is the **Sierra Morena**, a dark, austere-looking mountain range that is another of the plant-hunting regions mentioned by Polunin and Smythies (1991). They describe this range as having the largest expanse of maquis on the Iberian Peninsula. Although most of the area is rather stark, they recommend late May in the Despeñaperros Gorge, which cuts through the range with the main road and railroad line between Madrid and Córdoba, for its large variety of flowers.

Another of the outstanding plant destinations within easy reach of Córdoba and suitable for a late May visit is the **Sierra de Cazorla**, much of which has been protected as a national park since 1960. This rugged mountain range is at the northeastern extreme of Andalucía and has abundant rainfall of 80 inches (200 cm) per year. There are clear streams, deep ravines and extensive, relatively unspoiled oak and pine forests due to the area's former inaccessibility. Most of the lower-elevation vegetation is mediterranean, with a great abundance and variety of wildflowers. Quite a few species are ancient relicts that are endemic to the range. The parador, El Adelantado in the town of Cazorla, is an excellent base for exploring the park. The view from the inn reminds some visitors of the grandeur of Zion National Park and Bryce Canyon in southern Utah.

A few hours' drive to the southeast of Cazorla is the city of Granada, which lies against the snowy backdrop of the Sierra Nevada. Here is the most famous attraction of Andalucía,

FIG. 10.10. *Empty beaches in early spring on the rugged southern coast of Crete.*

the Moorish Alhambra, with its famous garden courtyards and fountains, the Court of the Myrtle Trees and the Court of the Lions.

GREECE: CRETE

A botanizing trip to Crete can also be enriched by visits to historic Minoan, Byzantine, Venetian, and Turkish sites. Crete is the first of the plant-hunting regions that Polunin (1997) describes in his field guide to Greece and the Balkans, and he devotes 19 detailed pages to suggested destinations with descriptions of what is likely to be in bloom and when. Many of the sites are at their best between late March and mid-May.

The island of Crete is about 30 by 150 miles (50 by 250 km), extending in an east-west direction. The largest towns and agricultural areas lie along the northern coast. The central spine of the island consists of a chain of limestone mountains with two peaks over 8,000 feet (2,400 m) high, Mount Ida (Mount Psiloritis) (Fig. 1.06) and Mount Díkti, on which snow may last into late May or beyond. The southern coast has rugged cliffs, beaches (Fig. 10.10), and narrow gorges that cut deeply into the mountains. The gorges provide shade and water for a special group of plant species that thrive under these conditions. Much of the cliffside vegetation is well preserved because of its inaccessibility and because it cannot be grazed.

Many of Crete's botanical attractions are in the wilder, less visited part of the island, west of the capital, **Iráklion** (note that place names are variously spelled when translated from the Greek). Just south of Iráklion is **Knossos**, the famous palace of King Minos, built in about 1700 B.C. and the best known archeological site in Crete. In the city itself is the noted Archeological Museum, devoted to the Minoan culture and containing many treasures excavated at Knossos and elsewhere on the island.

One of the plant-hunting excursions near Iráklion (Polunin, 1997) is to the top of Mount Youktas, above the village of Arkhanai, 13 miles (21 km) south of the city. The small pilgrimage Church of the Transfiguration is on the 2,650-foot (800 m) peak, and there are spectacular views to Iráklion and the sea to the north, Mount Psiloritis to the west, and Mount Díkti to the east. The peak is surrounded by large expanses of garrigue with a multitude of plant species, many of them rare pre-ice age relicts.

The other famous Minoan site on Crete is **Festós** (Fig. 10.11), which has not been as extensively reconstructed as Knossos. It has a beautiful setting on the south coast of Crete, south of Iráklion, overlooking the plain of Mesara and the sea. The trail over the hillsides between Festós and the village of Ayia Triada is only two miles (3 km) long and has abundant wildflowers, being particularly rich in orchids during April.

To the west of Iráklion are two smaller, north coast towns that can serve as a base, Rethymnon and Chania. Both have interesting, old town centers with buildings from the years of Venetian and Turkish rule. Chania and the surrounding area remain less visited than other parts of Crete because there are few sandy beaches nearby and it is further from the best known archeological sites. An alternative to Rethymnon and Chania are many smaller towns and villages, such as Spili, that provide a quieter, more relaxing atmosphere.

The **Samaria Gorge**, south of Chania, is one of the longest and most beautiful in Europe, extending from about 6,900 feet (2,100 m) down to sea level. It is a national park

and was declared a Biosphere Reserve in 1981. It is the most frequented of Greece's national parks, accounting for over 200,000 visitors out of a total of less than 500,000 (Allin, 1990). Most visitors come in the summer for the long one-day excursion. The gorge is not likely to be walkable before May. Organized tours drop visitors off at the top

FIG. 10.11. *Minoan ruins at Festós in Crete.*

of the gorge very early in the morning, with a boat pickup about six hours later at the bottom and a bus return in the evening.

The approach to the gorge makes a good day trip by itself since it is an excellent area for wildflowers. Above the village of Laki, the road winds up a steep hillside of phrygana, with orchids, bulbs, daphne, and cyclamen. On top is 3,300-foot (1,000 m) high Omalos plain, where April flowers include the deep pink *Tulipa saxatilis*.

Another of the many other gorges is the **Imbros Gorge**, which is the route taken by the road from Chania to the south coast village of Sfakia. Sfakia can be used a base for exploring the gorge, with its many endemic species. Another excursion from Sfakia is to the village of Anopolis, near which one can make the steep descent down the tulip- and orchid-rich Aradhena gorge.

Plakias, on the coast south of Rethymnon, is another possible base. Near the town itself is one of only five remaining groves of *Phoenix theophrasti*, the wild Cretan date palm mentioned by Theophrastus. Day trips from Plakias can be made to the small mountain towns of Spili and the villages of the Amari Valley that lie within an area rich in orchids and tulips.

ITALY

Italy has many options for nature travelers. Jepson's *Wild Italy* (1994) is useful for obtaining information about various regions and for a bibliography that includes regional wildflower books in Italian.

Tuscany and the Riviera di Levante. In this region, one can combine a visit to coastal areas of great natural beauty, such as the Maremma, with stays in the great Renaissance cities of Florence and Sienna. The western coast of Tuscany and the Italian Riviera south of Genoa (the Riviera di Levante) are highly developed. In the summer, this coast is among the most heavily visited areas of Europe, but in the spring it is much less crowded. Several easily accessible places have picturesque coastal wildflower walks.

The **Maremma** is a coastal nature reserve in the southern part of Tuscany, on the western side of the Italian peninsula. It occupies nine miles (15 km) of coastline, most of which is off limits to cars. The park headquarters is along its eastern boundary in the village of Alberese. This is where hiking maps and books about the park are sold and where there are regularly scheduled buses to the trailheads. There are several coastal trails that lead to beaches, coastal dunes, maquis, and forests of umbrella pine, all within a small area. Another trail climbs the wooded hills further inland. These are as high as 1,300 feet (400 m) and have splendid views over the coastal forests, beaches, and the sea. The park has the northernmost dwarf fan palms.

Further north along the coast, the Riviera di Levante has two areas that are known for their wildflowers and rugged coastal scenery. The coastal enclave of the **Cinque Terre**, or "five lands" consists of five villages linked by a coastal path, the Sentiero Azzuro, which courses through stands of Aleppo pines, maquis, and garrigue. Abundant wildflowers include several species of orchid. Sections of the walk between the village of Vernazza to Corniglia or in the other direction to Monterosso take one over lush, sunny headlands. Vernazza can be reached by a local railway from Portovenere or Levanto, where there are accommodations. Another trail follows the mountain ridge of the coast range, including ten peaks, the highest of which is 2,650 feet (800 m). Trail maps and guide books for the region are available in the local tourist offices and book stores.

A short distance south of Genoa, off the dramatic coastal expressway, is a quiet nature park around the 2,000-foot (600 m) high **Monte di Portofino**. The park is reached by trails from the picturesque harbor towns of Portofino and Camogli. Pine forests and coastal maquis are found on the south-facing slopes and a more central European, continental vegetation toward the north.

ALTERNATIVES IN SOUTHERN ITALY, SICILY, AND SARDINIA

For wilder and more remote areas, one has to travel to southern Italy, Sicily, and Sardinia, where some national parks have been established and many sites have been proposed, but are as yet largely undeveloped (Jepson, 1994). **Monti del Gennargentu** in Sardinia has some of the Mediterranean Basin's largest and most ancient forests of holm oak, gorges filled with oleander, and extensive upland plateau areas of maquis. The **Parco Nazionale della Calabria** protects three separate parts of the Sila Mountains, an area that remains as wild as any on the Italian peninsula and is to some extent still bandit-ridden like other parts of the south. There is a visitors' center at Silla Grande, in one of the three sub-ranges.

Other parks and reserves in the south are more developed for visitors. In Sicily, the area around the active, snow-covered volcano, **Mount Etna**, has a multitude of marked trails and ample tourist accommodations at its base. It has coastal maquis on its lower slopes, vineyards and orchards higher up, followed by forests of holm oak, chestnut, beech, and Corsican pine. Another easily reached area is the **Gargano and Foresta Umbra**, on the eastern coast of the Italian peninsula. It is interesting for its abundance of plant species that are otherwise more typical of the Balkans, such as *Campanula garganica* and *Ranunculus garganicus*. The town of Péschici, on the Adriatic Sea north of the park, is a good base for exploring this area, which is a half-day trip by expressway from Rome or Naples.

REFERENCES

Acocks JPH. (1988) *Veld Types of South Africa.* Reprinted. Government Printer, Pretoria.

Allin CW. (1990) *International Handbook of National Parks and Nature Reserves.* Greenwood Press, New York.

Arroyo MTK, Cavieres L, Marticorena C, Muños-Schick M. (1995) Convergence in the mediterranean floras in Central Chile and California: insights from comparative biogeography. In *Ecology and Biogeography of Mediterranean Ecosystems in Chile, California, and Australia.*, Arroyo MTK, Zedler PH, Fox MD, eds. Springer Verlag, New York. pp 43-88.

Arroyo MTK, Zedler PH, Fox MD. (1995) *Ecology and Biogeography of Mediterranean Ecosystems in Chile, California, and Australia.* Springer Verlag, New York.

Aschmann H. (1973) Distribution and peculiarity of mediterranean ecosystems. in *Mediterranean Type Ecosystems. Origin and Structure.* DiCastri F and Mooney HA, eds. Springer Verlag, New York.

Aschmann H. (1977) Man's impact on the several regions with mediterranean climates. in *Convergent Evolution in Chile and California.* Mooney, HA, ed., Dowden, Hutchinson & Ross, Stroudsburg, pp 363-371.

AUSLIG (Australian Surveying and Land Information Group) (1990). *Atlas of Australian Resources. Volume 6, Vegetation.* Commonwealth Government Printer, Canberra.

Axelrod DI. (1973) History of the mediterranean ecosystem in California. In *Mediterranean Type Ecosystems. Origin and Structure.* DiCastri F, Mooney HA., eds. Springer, New York.

Axelrod DI. (1989) Age and origin of chaparral. In *The California Chaparral: Paradigms Reexamined.* Keeley SC, ed. Natural History Museum of Los Angeles County, Publ No 34. pp. 7-19.

Bagnouls F, Gaussen H. (1957) Les climats écologiques et leur classification. Annls Geogr 66:193-220.

Bakker E. (1984) *An Island Called California. An Ecological Introduction to Its Natural Communities.* 2nd edition. University of California Press, Berkeley.

Barbour MG, Minnich RA. (1990) The myth of chaparral convergence. Israel J Bot 39:453-463.

Barbour MG, Major, J, eds. (1988) *Terrestrial Vegetation of California.* California Native Plant Society, Sacramento.

Barbour MG, Pavlik B, Drysdale F, Lindstrom S. (1993) *California's Changing Landscapes.* California Native Plant Society, Sacramento.

Barlow BA. (1994) Phytogeography of the Australian region. In *Australian Vegetation.* 2nd ed, Groves RH, ed, Cambridge University Press, Cambridge, Great Britain. pp 3-35.

Baumann H. (1993) *The Greek Plant World in Myth, Art, and Literature.* Timber Press, Portland.

Bean WJ. (1976) *Trees and Shrubs Hardy in the British Isles.* 8th edition, Clarke DL, Taylor G, eds. Murray, London.

Beard JS. (1969) Endemism in the Western Australian flora at the species level. J Roy Soc West Australia 52(1):18-20.

Beard JS. (1990) *Plant Life of Western Australia.* Kangaroo Press, Kenthurst.

Birot P, Dresch J. (1953) *La Méditerranée et le Moyen-Orient, Tome Premier: La Méditerranée Occidentale*. *Géographie Physique et Humaine*. Presses Universitaires de France, Paris.

Blamey M, Grey-Wilson D. (1993) *Mediterranean Wild Flowers*. Domino Books, St Helier, Jersey.

Blanchard JW (1990) *Narcissus. A Guide to Wild Daffodils*. Alpine Garden Society, Woking Surrey.

Blondel J, Aronson J. (1995) Biodiversity and ecosystem function in the Mediterranean Basin: human and non-human determinants. In: *Mediterranean-Type Ecosystems. The Function of Biodiversity*. Davis GW, Richardson DM, eds. Springer, Berlin. pp 43-120.

Bond P, Goldblatt P. (1984) Plants of the Cape flora. A descriptive catalog. J South African Botany. Suppl 13.

Boomsma CD. (1972) *Native Trees of South Australia*. Woods and Forests Department, South Australia.

Briggs JD, Leigh JH. (1988) *Rare or Threatened Australian Plants*. special publication 14, ANPWS, Canberra.

Brun D, Lobréaux O, Maistre M, Perret P, Romane F. (1990) Germination of *Quercus ilex* and *Q. pubescens* in a *Q. ilex* coppice. Long-term consequences. Vegetatio 87:45-50.

Bryan JE. (1989) *Bulbs, Volume I*. Timber Press, Portland.

Burton R. (1991) *Nature's Last Strongholds*. Oxford University Press, New York.

Butterfield HM. (1963) *A History of Subtropical Fruits and Nuts in California*. University of California, Division of Agricultural Sciences.

Camus A (1934-39) *Les Chênes. Monographie du genre Quercus*. Lechevalier, Paris.

Canadell J, Zedler PH. (1995) Underground structures of woody plants in mediterranean ecosystems of Australia, California, and Chile. In *Ecology and Biogeography of Mediterranean Ecosystems in Chile, California, and Australia*, Arroyo MTK, Zedler PH, Fox MD, eds. Springer Verlag, New York. pp 177-210.

CNC (California Nature Conservancy). (1987) *Sliding Towards Extinction: The State of California's Natural Heritage*. San Francisco.

Cody ML. (1986) Diversity, rarity, and conservation in mediterranean-climate regions. In *Conservation Biology. The Science of Scarcity and Diversity*. Soulé ME, ed. Sinauer, Sunderland, MA.

Cody ML, Mooney HA. (1978) Convergence versus nonconvergence in mediterranean-climate ecosystems. Ann Rev Ecol Syst 9:265-321.

Cowling RM. (1992) *The Ecology of Fynbos. Nutrients, Fire and Diversity*. Oxford University Press, Cape Town.

Cowling RM, Holmes PM. (1992) Flora and vegetation. In *The Ecology of Fynbos. Nutrients, Fire and Diversity*. Oxford University Press, Cape Town. pp 23-61.

Cowling RM, Richardson D. (1995) *Fynbos. South Africa's Unique Floral Kingdom*. Fernwood Press, Vlaeberg, South Africa.

Cowling RM, Rundel PW, Lamont, BB, Arroyo MK, Arianoutsou M. (1996) Plant diversity in mediterranean-climate regions. Trends in Ecology and Evolution 11:362-366.

Daget P. (1980) *Atlas d'Aréologie Périméditerranée*. Institut de Botanique. Montpellier.

Dashorst GRM, Jessop JP. (1990) *Plants of the Adelaide Plains and Hills*. Kangaroo Press, Kenthurst, NSW.

Davis SD et al. (1986) *Plants in Danger. What Do We Know?* IUCN, Gland.

Deacon HC. (1983) The comparative evolution of mediterranean-type ecosystems: a southern perspective. In: *Mediterranean-Type Ecostystems. The Role of Nutrients*. Kruger FJ, Mitchell DT, and Jarvis JUM, eds. Springer Verlag, Berlin. pp 3-40.

DeBano LF, Dunn PH, Conrad CE. (1977) Fire's effect on physical and chemical properties of chaparral soils. In *Symposium on the Environmental Consequences of Fire and Fuel Management in the Mediterranean Ecosystems*. USDA Forest Service Technical Report WO-3. U.S. Department of Agriculture, Washington, D.C.

Debazac EF. (1983) Temperate broad-leaved evergreen forests of the Mediterranean region and Middle East. In *Temperate Broad-Leaved Evergreen Forests*, Ovington JD, ed, Elsevier, Amsterdam. pp 107-123.

Deghi GS, Huffman T, Culver JW. (1995) California's native Monterey pine populations: potential for survival. Fremontia 23:14-23.

DeSimone S. (1995) California's coastal sage scrub. Fremontia 23(4):3-8.

Dewitt JB. (1993) *California Redwood Parks and Preserves*. Save-the-Redwoods League. San Francisco.

DiCastri F, Mooney HA. (1973) *Mediterranean Type Ecosystems. Origin and Structure*. Springer, New York.

DiCastri F, Goodall DW, Specht RL., eds. (1981) *Ecosystems of the World II. Mediterranean-Type Shrublands*. Elsevier, Amsterdam.

DiCastri F. (1991) An ecological overview of the five regions with a mediterranean climate. In *Biogeography of Mediterranean Invasions*. Groves RH, Di Castri F, eds. Cambridge University Press, Cambridge. pp 3-16.

Dodson JR. (1994) Quaternary vegetation history. In *Australian Vegetation*. 2nd ed, RH Groves, ed, Cambridge University Press. Cambridge, Great Britain. pp 37-56.

Donley MW, Allan S, Caro P, Patton CP. (1979) *Atlas of California*. Culver City, Pacific Book Center.

Doutt R. (1994) *Cape Bulbs*. Timber Press, Portland.

Dudal R, Tavernier R, Osmond D. (1966) Soil map of Europe, 1:2,500,000 (with climate maps). European Commission on Agriculture. Food and Agriculture Organization of the United Nations. Rome.

Dunn AT. (1986) Fire history in San Diego County. Fremontia 14(3):24-27.

Dunn AT. (1988) The biogeography of the California Floristic Province. Fremontia 15(4):3-9.

Emberger L. (1955) Project d'une classification biogeographique des climats. In *Les Divisions Ecologiques du Monde*. Centre de Recherche Scientifique, Paris. pp 5-11.

Erickson R, George AS, Marchant NG, Morcombe MK. (1973) *Flowers and Plants of Western Australia*. Reed, Sydney.

Ern H. (1966) Die dreidimensionale Anordnung der Gebirgsvegetation auf der Iberischen Halbinsel. Bonn Geogr Abh 37:1-132.

Ewing RA (1990) How are oaks protected? What are the issues? Fremontia 18(3): 83-88.

Faber P. ed, (1990) *The Year of the Oak*. Fremontia 18(3):1-112.

Fairall AR. (1970) *West Australian Plants in Cultivation*. Pergamon Press, Rushcutters Bay, Australia.

Farjon A. (1984) *Pines*. Brill/Backhuys, Leiden. 1984.

Fox MD. (1995) Australian mediterranean vegetation: intra- and intercontinental comparisons. In *Ecology and Biogeography of Mediterranean Ecosystems in Chile, California, and Australia*. Arroyo MTK, Zedler PH, Fox MD, eds. Springer Verlag, New York. pp 43-88.

Frome M, Wauer RW, Pritchard PC. (1990) United States: National Parks. In *International Handbook of National Parks and Nature Reserves*. Allin CW, ed. Greenwood Press, New York.

Fuentes ER. (1988) Sinopsis de paisajes de Chile central. In *Ecologia del Paisaje en Chile Central. Estudios Sobre sus Espacios Montañosos*. Fuentes E, Prenafeta S, eds. Ediciones Universidad Católica de Chile, Santiago.

Fuentes ER. (1983) Defoliation patterns in matorral ecosystems. In: *Mediterranean-Type Ecostystems. The Role of Nutrients*. Kruger FJ, Mitchell DT, and Jarvis JUM, eds. Springer Verlag, Berlin. pp 525-542.

Fuentes ER, Hoffmann AJ, Poiani A, Alliende MC. (1986) Vegetation change in large clearings: patterns in the Chilean matorral. Oecologia 68:358-366.

Fuentes ER, Segura AM, Holmgren M. (1994) Are the responses of matorral shrubs different from those of an ecosystem with a reputed fire history? In: *The Role of Fire in Mediterranean-Type Ecosystems*. Moreno JM, Oechel WC, eds. Springer Verlag, New York. pp17-25.

Fuentes ER, Muñoz MR. (1995) The human role in changing landscapes in Central Chile: implications for intercontinental comparisons. In *Ecology and Biogeography of Mediterranean Ecosystems in Chile, California, and Australia*, Arroyo MTK, Zedler PH, Fox MD, eds. Springer Verlag, New York. pp 401-417.

Fuentes ER, Montenegro G, Rundel PW, Arroyo MTK, Ginocchio R, Jaksic FM. (1995) Functional approaches to biodiversity in the mediterranean-type ecosystems of Central Chile. In *Mediterranean-type Ecosystems. The Function of Biodiversity*. Davis GW, Richardson DM, eds. Springer, Berlin.

George AS. (1986) *The Banksia Book*. Kangaroo Press, Kenthurst, NSW.

George AS, Hopkins AJM, Marchant NG. (1979) The heathlands of Western Australia. In *Heathlands and Related Scrublands of the World*. Specht RL, ed. Elsevier, Amsterdam.

Goldblatt P. (1978) An analysis of the flora of Southern Africa: its characteristics, relationships, and origins. Ann Missouri Bot Gardens 65:369-436.

Gomez-Campo C. (1985) The conservation of Mediterranean plants: principles and problems. In *Plant Conservation in the Mediterranean Area*. Junk, Dordrecht.

Good R. (1964) *The Geography of Flowering Plants* 3rd edition. Longman, London.

Grenon M, Batisse M, eds. (1989) *Futures for the Mediterranean Basin. The Blue Plan*. Oxford University Press, Oxford.

Greuter W. (1991) Botanical diversity, endemism, rarity, and extinction in the mediterranean area: an analysis based on the published volumes of Med-Checklist. Bot Chron 10:63-79.

Griffin JR. (1988) Oak Woodland. In *Terrestrial Vegetation of California*. Barbour MG, Major J, eds. California Native Plant Society. pp 383-415.

Griffin JR, Muick PC. (1990) California native oaks: past and present. Fremontia 18:4-11.

Griffin T, McCaskill M. (1986) Atlas of South Australia. South Australia Government Printing Division. p 51.

Groves RH. (1991) The biogeography of mediterranean plant invasions. In *Biogeography of Mediterranean Invasions*. Groves RH, DiCastri F, eds. Cambridge University Press, Cambridge. pp 427-438.

Grunfeld FV. (1994) *Wild Spain*. Sierra Club Books, San Francisco.

Hanes TL. (1988) California Chaparral. In *Terrestrial Vegetation of California*. Barbour MG, Major J, eds. California Native Plant Society. pp 417-469.

Harant H, Jarry D. (1971) The thickets and woods of the Mediterranean region. In *World Vegetation Types*. Eyre SR, ed. Columbia University Press, New York. pp174-185.

Hawkins L. (1990) Gardener's guide to the mediterranean climate. In *The Pacific Horticulture Book of Western Gardening*. Waters G, Harlow N, eds., Boston, Godine. pp 3-9.

Heywood VH. (1978) *Flowering Plants of the World*. Mayflower Books, New York.

Hilton-Taylor C, A Le Roux. (1989) Conservation status of the fynbos and karoo biomes. In *Biotic Diversity in Southern Africa*. Huntley BJ, ed. Oxford University Press, Cape Town.

Hobbs RJ, Richardson DM, Davis GW. (1995) Mediterranean-type ecosystems: opportunities and constraints for studying the function of biodiversity. In: *Mediterranean-Type Ecosystems. The Function of Biodiversity*. Davis GW, Richardson DM, eds. Springer, Berlin. pp 1-42.

Hoffmann AJ, Armesto JJ. (1995) Modes of seed dispersal in the mediterranean regions in Chile, California, and Australia. In *Ecology and Biogeography of Mediterranean Ecosystems in Chile, California, and Australia*, Arroyo MTK, Zedler PH, Fox MD, eds. Springer Verlag, New York. pp 289-310.

Hoffmann AJ. (1995) *Flora Silvestre de Chile, Zona Central*. 3rd ed. Santiago, Fundacion Claudio Gay.

Hoffmann AJ. (1991) *Flora Silvestre de Chile, Zona Araucana*, 2nd ed. Santiago, Fundacion Claudio Gay.

Hoffmann AJ, Fuentes ER. (1988) ¿Es necesario conservar? In *Ecologia del Paisaje en Chile Central*. Fuentes E, Prenafeta S, eds. Ediciones Universidad Católica de Chile, Santiago.

Hoffmann JAJ. (1975) *Climatic Atlas of South America*. OMM.WHO. UNESCO. Cartographia, Budapest.

Holing D. (1988) *California Wild Lands. A Guide to Nature Conservancy Preserves*. Chronicle Books, San Francisco.

Hölldobler B, Wilson EO. (1990) *Ants*. Harvard University Press, Cambridge.

Holliday I, Overton B, Overton D. (1994) *Kangaroo Island's Native Plants*. Swift Printing, Adelaide.

Hooker, JD. (1853) *The Botany of the Antarctic Voyage of H. M. Discovery Ships Erebus and Terror in the Years 1839-1843*. Lovell Reeve, London.

Hopper SD. (1992) Patterns of plant diversity at the population and species levels in south-west Australian mediterranean ecosystems. In *Biodiversity of Mediterranean Ecosystems in Australia*. Hobbs RJ, eds. Surrey Beatty & Sons, Sydney.

Hopper SD, Harvey MS, Chappill JA, Main AR, York-Main, B. (1996) The Western Australian biota as Gondwanan heritage - a review. In *Gondwanan Heritage: Past, Present and Future of the Western Australian Biota*. Hopper SD, Chappill JA, Harvey MS, George AS, eds. Surrey Beatty & Sons, Chipping Norton.

Huntley BJ, ed. (1989) *Biotic Diversity in Southern Africa: Concepts and Conservation*. Oxford University Press, Cape Town.

Huttary J. (1950) Die Verteilung der Niederschläge auf die Jahreszeiten im Mittelmehrgebiet. Meteorol Rdsch 3:111-119.

Huxley A, Taylor W. (1989) *Flowers of Greece and the Aegean*. Hogarth Press, London.

Huxley L. (1918) *Life and Letters of Sir Joseph Dalton Hooker*. Appleton & Company, New York.

Instituto Geografico Militar. (1983). *Atlas de la Republica de Chile*. Instituto Geografico Militar, Santiago.

James PE. (1965) *A Geography of Man*. Ginn, Boston.

Jeppe B, Duncan G. (1989) *Spring and Winter Flowering Bulbs of the Cape*. Oxford University Press, Cape Town.

Jepson T. (1994) *Wild Italy*. Sierra Club Books, San Francisco.

Johnson SD. (1992) Plant-animal relationships. In *The Ecology of Fynbos*, Cowling RM, ed. Oxford University Press, Cape Town. pp 175-205.

Johnston VR. (1994) *California Forests and Woodlands*. University of California Press, Berkeley.

Jung W, Selmeier A, Dernbach. (1992). *Araucaria*. D'Oro Verlag, Lorsch, Germany.

Kassioumis K. (1990) Greece. In *International Handbook of National Parks and Nature Reserves*. Allin, CW, ed., Greenwood Press, New York.

Keeley JE. (1992) A Californian's view of fynbos. In *The Ecology of Fynbos*, Cowling RM, ed. Oxford University Press, Cape Town. pp 372-387.

Keeley JE. (1995) Seed-germination patterns in fire-prone mediterranean-climate regions. In *Ecology and Biogeography of Mediterranean Ecosystems in Chile, California, and Australia*, Arroyo MTK, Zedler PH, Fox MD, eds. Springer Verlag, New York. pp 239-273.

Keeley JE, Keeley SC. (1986) Chaparral and wildfires. Fremontia 14:18-21.

Kellogg A. (1889) *Illustrations of West American Oaks*. San Francisco.

Knight R. (1991) West coast strandveld; a paradise lost? Veld & Flora, December, pp 100-103.

Köhlein F. (1987) *Iris*. Timber Press, Portland.

Kreissman B. (1991) *California. An Environmental Atlas and Guide*. Bear Klaw Press, Davis.

Kruger FJ. (1982) Prescribing fire frequencies in Cape fynbos in relation to plant demography. Gen Tech Rep PSW-58. Berkeley, CA: Pacific Southwerst Forest and Range Experiment Station, Forest Service, U.S. Department of Agriculture. pp 118-122.

Küchler A W. (1977) *Natural Vegetation of California*. A map in *Terrestrial Vegetation of California*. Barbour MG, Major J, eds. (1977) California Native Plant Society, Sacramento.

Kummerow J. (1981) Structure of roots and root systems. In *Ecosystems of the World II. Mediterranean-Type Shrublands*. DiCastri F, Goodall DW, Specht RL., eds. Elsevier, Amsterdam. pp 269-288.

Kummerow J, Borth W. (1986) Mycorrhizal associations in chaparral. Fremontia 14(3);11-13.

Lalande P. (c1968) Vegetation map of the Mediterranean region, scale 1;5,000,000. Eastern sheet by Lalande P, western sheet by Bagnouls F, Gaussen H. Drawn by Garcia H, Pivot H. UNESCO, New York.

Lamont BB (1983) Strategies for maximizing nutrient uptake in two mediterranean ecosystems of low nutrient status. In: *Mediterranean-Type Ecostytems. The Role of Nutrients*. Kruger FJ, Mitchell DT, and Jarvis JUM, eds. Springer Verlag, Berlin. pp 247-273.

Le Maitre DC, Midgley JJ. (1992) Plant reproductive ecology. In *The Ecology of Fynbos*. Cowling RM, ed. Oxford University Press, Cape Town. pp 135-174.

Lenz LW, Dourley J. (1981) *California Native Trees and Shrubs*. Rancho Santa Ana Botanic Garden, Claremont.

Lepart J, Debussche M. (1992) Human impact on landscape patterning: Mediterranean examples. In *Landscape Boundaries. Consequences for Biotic Diversity and Ecological Flows*. Hansen AJ and DiCastri F, eds. Springer, New York. pp 76-106.

Lighton C. (1973) *Cape Floral Kingdom*. Juta and Co., Cape Town.

Linder HP, Meadows ME, Cowling RM. (1992) History of the Cape Flora. In *The Ecology of Fynbos*. Cowling RM, ed. Oxford University Press, Cape Town. pp 113-134.

Lyons K, Cooney-Lazaneo MB. (1988) *Plants of the Coast Redwood Region*. Looking Press, Boulder Creek.

Mabberly DJ, Placito PJ. (1993) *Algarve Plants and Landscapes, Passing Traditions and Ecological Change*. Oxford University Press, Oxford.

Major J. (1988) California climate in relation to vegetation. In *Terrestrial Vegetation of California*. Barbour MG, Major J, eds. California Native Plant Society, Sacramento. pp 11-74.

Margaris NS. (1981) Adaptive strategies in plants dominating mediterranean-type ecosystems. In *Ecosystems of the World II. Mediterranean-Type Shrublands*. DiCastri F, Goodall DW, Specht RL., eds. Elsevier, Amsterdam. pp309-315.

Mason H, DuPlessis E. (1972) *Western Cape Sandveld Flowers*. Struik, Cape Town.

Mathew B. (1982) *The Crocus*. Timber Press, Portland.

Mathews LJ, Carter Z. (1993) *Proteas of the World*. David Bateman, Aukland.

McClintock E. (1976) *Trees of Golden Gate Park*. A continuing series in Pacific Horticulture begun in 1976.

McClintock E. (1987) The displacement of native plants by exotics. In *Conservation and Management of Rare and Endangered Plants. Proceedings of a California Conference on the Conservation and Management of Rare and Endangered Plants*. Elias TS, ed. California Native Plant Society, Sacramento.

McCreary DD. (1990) Native oaks-the next generation. Fremontia 18(3):44-47.

McMahon L, Frazer M. (1988) *A Fynbos Year*. Philip, Cape Town.

Merlo ME, Alemán MM, Cabello J, Peñas J. (1993) On the Mediterranean fan palm (*Chamaerops humilis*). Principes 37:151-158.

Montenegro G, Rivera O, Bas F. (1978) Herbaceous vegetation in the Chilean matorral. Oecologia 36:237-244.

Montenegro, GR. (1984) *Atlas de Anatomía de Especies Vegetales Autóctonas de la Zona Central*. Ediciones Universidad Católica de Chile, Santiago.

Mooney HA, ed. (1977) *Convergent Evolution in Chile and California. Mediterranean Climate Ecosystems*. Dowden, Hutchinson, & Ross, Stroudsburg, PA.

Mooney HA. (1988) Lessons from Mediterranean-climate regions. In *Biodiversity* Wilson EO, Peter FM, eds. National Academy Press, Washington, D.C. pp 157-165.

Morandini, R. (1994) Proceedings of the workshop, *Improvement of Coppice Forests in the Mediterranean Region*. Ann Ist Sper Selv 23:257-333.

Morcombe, MK. (1968) *Australia's Western Wildflowers*. Morcombe, Armadale, Western Australia.

Morcombe, MK. (1970) *Australia's Wildflowers*. Lansdowne, Melbourne.

Mosley JG. (1990) Australia. In *International Handbook of National Parks and Nature Reserves*. Allin, CW, ed., Greenwood Press, New York.

Mulcahy MJ, White BJ, Havel JJ. (1988) Parks and reserves in the south-west forests. In *Case Studies in Environmental Hope*, Newman P, Neville S, Duxbury L, eds. Picton, Western Australia.

Muñoz MR, Fuentes ER. (1989) Does fire induce shrub germination in the Chilean matorral? Oikos 56:177-181.

Munz PA. (1968) California plant communities. In *A California Flora and Supplement*. University of California Press, Berkeley. pp 10-18.

Nash RF. (1990) *American Environmentalism. Readings in Conservation History*. 3rd edition. McGraw-Hill, New York.

NATMAP (Division of National Mapping). (1986). *Atlas of Australian Resources. Volume 4. Climate*. Commonwealth Government Printer. Canberra.

Naveh Z, Whittaker RH. (1979) Structural and floristic diversity of shrublands and woodlands in northern Israel and other Mediterranean areas. Vegetatio 43:171-190.

O'Leary JF. (1988) Habitat differentiation among herbs in postburn California chaparral and coastal sage scrub. Am Midland Naturalist 120:41-49.

O'Leary JF. (1995) Coastal sage scrub: threats and current status. Fremontia 23(4):27-31.

Ornduff R. (1974) *Introduction to California Plant Life*. University of California Press, Berkeley.

Orshan G, Floret CH, Le Floc'h E, Le Roux A, Montenegro G, Romane F. (1989) General synthesis. In *Plant Pheno-Morphological Studies in Mediterranean Type Ecosystems*. Orshan G, ed. Kluwer, Dortrecht. pp 389-404.

Parsons RF. (1994) Eucalyptus scrubs and shrublands. In *Australian Vegetation*. 2nd ed, Groves RH, ed, Cambridge University Press, Cambridge. pp 291-319.

Pate JS, Beard JS. (1984) *Kwongan. Plant Life of the Sandplain. Biology of a South-west Australian Shrubland Ecosystem*. University of Western Australia Press, Nedlands.

Pavlik BM, Muick PC, Johnson S, Popper M. (1991) *Oaks of California*. Cachuma Press, Los Olivos, CA.

Peattie DC. (1953) *A Natural History of Western Trees*. Bonanza Books, New York.

Polunin O, Huxley A. (1965) *Flowers of the Mediterranean*. Chatto & Windus, London.

Polunin O, Smythies BE. (1991) *Flowers of South-West Europe: A Field Guide*. Oxford.

Polunin O. (1997) *Flowers of Greece and the Balkans: A Field Guide*. Oxford University Press.

Pons A, Quézel P. (1985) The history of the flora and vegetation and past and present human disturbance in the Mediterranean region. In *Plant Conservation in the Mediterranean Area*. Gómez-Campo C, ed. Junk Publishers, Dortrecht. pp 25-43.

Pryor LD. (1976) *The Biology of Eucalypts*. Camelot Press, Southampton.

Quézel P. (1981) Floristic composition and phytosociological structure of sclerophyllous matorral around the Mediterranean. In *Ecosystems of the World II. Mediterranean-Type Shrublands*. DiCastri F, Goodall DW, Specht RL, eds. Elsevier, Amsterdam. pp 107-121.

Quinn RD. (1986) Mammalian herbivory and resilience in mediterranean-climate ecosystems. In: *Resilience in Mediterranean-type Ecosystems*. Dell B, Hopkins AJM, Lamont BB, eds. Junk Publishers, Dortrecht. pp 113-128.

Rackham O. (1995) *The History of the Countryside*. Weidenfeld and Nicolson, London.

Ràfols ES. (1977) Analog tree and shrub species. In *Chile California Scrub Atlas*. Thrower NJW, Bradbury DE. eds. Dowden, Hutchinson & Ross, Stroudsburg, PA. pp 129-143.

Ràfols ES, Kummerow J. (1977) Vegetation transects. In (1979) *Chile California Scrub Atlas*. Thrower NJW, Bradbury DE. eds. Dowden, Hutchinson & Ross, Stroudsburg, PA. pp 83-91.

Raine P. (1990) *Mediterranean Wildlife. The Rough Guide*. Harrap-Columbus, London.

Ramírez C, Hauenstein E, San Martín J, Contreras D. (1989) Study of the flora of Rucamanque, Cautin Province, Chile. Ann Missouri Botanical Garden 76:444-453.

Raven PH. (1973) The evolution of mediterranean floras. In *Mediterranean Type Ecosystems. Origin and Structure*. DiCastri F, Mooney HA, eds. Springer, New York.

Raven PH, Axelrod DI. (1978) *Origin and Relationships of the California Flora*. University of California Press, Berkeley.

Raven PH. (1988) The California flora. In *Terrestrial Vegetation of California*. Barbour MG, Major J, eds. California Native Plant Society, Sacramento. pp 11-74.

Read DJ, Mitchell DT. (1983) Decomposition and mineralization processes in mediterranean-type ecosystems and in heathlands of similar structure. In: *Mediterranean-Type Ecostystems. The Role of Nutrients*. Kruger FJ, Mitchell DT, Jarvis JUM, eds. Springer Verlag, Berlin. pp 208-232.

Richardson DM, Macdonald IAW, Holmes PM, Cowling RM. (1992) Plant and animal invasions. In: *The Ecology of Fynbos*. Cowling RM, ed. Oxford University Press, Cape Town.

Robinson AC. (1985) Island management in South Australia. In *Australian and New Zealand Islands: Nature Conservation Values and Management*. Burbridge A, ed. Proceedings of a Technical Workshop, Barrow Island, Western Australia, 1985. Department of Conservation and Land Management.

Rother K. (1984) *Die mediterranen Subtropen* Höller & Zwick. Braunschweig. New printing: 1993. Westermann, Braunschweig.

Rottmann J. (1988) *Bosques de Chile. Chile's Woodlands.* Ograma, Santiago.

Rourke JP, Anderson F. (1982) *The Proteas of Southern Africa.* Centaur Publishers, Johannesburg.

Rundel, PW. (1979) Ecological impact of fires on mineral and sediment pools and fluxes. In *Fire and Fuel Management in Mediterranean-Climate Ecosystems: Research Priorities and Programmes.* Agee JK, ed., MAB Technical Note 11. UNESCO, Paris, pp 17-21.

Rundel, PW. (1981) The matorral zone of Central Chile. In *Ecosystems of the World II. Mediterranean-Type Shrublands.* DiCastri F, Goodall DW, Specht RL., eds. Elsevier, Amsterdam. pp 175-201.

Rundel, PW. (1986) Structure and function in California chaparral. Fremontia 14(3):3-10. (issue devoted to chaparral)

Rundel PW, Weisser PJ. (1975) La Campana, a new national park in Central Chile. Biol Conserv 8:35-46.

Sampson AW (1944) Plant succession on burned chaparral lands in Northern California. Agricultural Experiment Station. Bulletin 685. University of California College of Agriculture, Berkeley. pp 144.

Sauer JD. (1988) *Plant Migration. The Dynamics of Geographic Patterning in Seed Plant Species.* University of California Press, Berkeley.

Sawyer JO, Keeler-Wolf T. (1995) *A Manual of California Vegetation.* California Native Plant Society, Sacramento.

Schick, MM. (1991) *Flores del Norte Chico.* 2nd ed. Direccion de Bibliotecas Archivos y Museos, Santiago.

Schmida A, Barbour M. (1982) A comparison of two types of mediterranean scrub in Israel and California. Gen Tech Rep PSW-58. Berkeley, CA: Pacific Southwest Forest and Range Experiment Station, Forest Service, U.S. Department of Agriculture. pp 100-106.

Schmithüsen J. (1956) Die räumliche Ordnung der chilenischen Vegetation. Bonn Geogr Abh 17:1-89.

Schoenherr AA. (1992) *A Natural History of California.* University of California Press, Berkeley.

Schönfelder I, Schönfelder P. (1990) *Wild Flowers of the Mediterranean.* Collins, London.

Schumann D, Kirsten G, Oliver EGH. (1992) *Ericas of South Africa.* Fernwood Press, Vlaeberg, South Africa.

Scott TA. (1990) Conserving California's rarest white oak: the Engelmann oak. Fremontia 18(3):26-29.

Serrada R, Allué M, San Miguel A. (1994) The coppice system in Spain. Current situation, state of art and major areas to be investigated. Ann Ist Sper Selv 23:266-275.

Smith AG, Hurley AM, Briden JC. (1981) *Phanerozoic Paleocontinental World Maps.* Cambridge Earth Science Series. Cambridge University Press, Cambridge.

Smuts R, Bond B. (1995) Namaqualand's carpets of colour conceal an insect desert. Veld & Flora 81:70-73.

Sparrow A. (1989) Mallee vegetation in South Australia. In *Mediterranean Landscapes in Australia. Mallee Ecosystems and their Management.* Noble JC, Bradstock RA, eds. CSIRO, Australia.

Specht RL. (1972) *The Vegetation of South Australia.* 2nd edition. AB James, Government Printer, Adelaide.

Specht RL. (1981) Conservation: Australian Heathlands. In *Ecosystems of the World. Vol. 9B. Heathlands and Related Shrublands.* Specht RL, ed. Elsevier, Amsterdam.

Specht RL. (1981) Mallee ecosystems in Southern Australia. In *Ecosystems of the World II. Mediterranean-Type Shrublands.* DiCastri F, Goodall DW, Specht RL., eds. Elsevier, Amsterdam. pp 203-231.

Specht RL. (1982) General characteristics of mediterranean-type ecosystems. Gen Tech Rep PSW-58. Berkeley, CA: Pacific Southwerst Forest and Range Experiment Station, Forest Service, U.S. Department of Agriculture. pp 13-19.

Specht RL. (1994) Heathlands. In *Australian Vegetation*. 2nd ed, Groves RH, ed, Cambridge University Press. Cambridge. pp 321-344.

Specht RL, Moll EJ. (19⌐3) Mediterranean-type heathlands and sclerophyllous shrublands of the world: an overview. In: *Mediterranean-Type Ecosystems. The Role of Nutrients*. Kruger FJ, Mitchell DT, and Jarvis JUM, eds. Springer Verlag, Berlin. pp 41-65.

Stankey GH. (1990) United States: Wilderness Areas. In *International Handbook of National Parks and Nature Reserves*. Allin CW, ed., Greenwood Press, New York.

Steward D, Webber PJ (1981) The plant communities and their environments. In *Resource Use by Chaparral and Matorral. A Comparison of Vegetation Function in Two Mediterranean Type Ecosystems*. Miller PC, ed. Springer, New York. pp 43-68.

Sweeney JR. (1956) *Responses of Vegetation to Fire. A Study of the Herbaceous Vegetation Following Fires*. University of California Publications in Botany 28:143-250. University of California Press.

Sykes WR, Godley EJ. (1968) Transoceanic dispersal of *Sophora* and other genera. Nature 218 (5140):495-496.

Takhtajan A. (1986) *Floristic Regions of the World*. University of California Press, Berkeley.

Talbot AM. (1960) *Atlas of the Union of South Africa*. Government Printer, Pretoria.

Taylor J. (1992) *The Living West of Australia*. Kangaroo Press, Kenthurst, NSW.

Thomas B. (1981) *The Evolution of Plants and Flowers*. St. Martin's Press, New York.

Thrower NJW, Bradbury DE. (1973) The physiography of the mediterranean lands with special emphasis on California and Chile. In *Mediterranean-Type Ecosystems. Origin and Structure*. DiCastri F and Mooney HA, eds. Springer Verlag, New York.

Thrower NJW, Bradbury DE. (1979) *Chile California Scrub Atlas*. Dowden, Hutchinson & Ross, Stroudsburg, PA.

Tomaselli R. (1981) Main physiognomic types and geographic distribution of shrub systems related to mediterranean climates. In *Ecosystems of the World II. Mediterranean-Type Shrublands*. DiCastri F, Goodall DW, Specht RL., eds. Elsevier, Amsterdam. pp 95-106.

Trabaud L. (1994) Postfire plant community dynamics in the Mediterranean Basin. In *The Role of Fire in Mediterranean-Type Ecosystems*. Moreno JM, Oechel WC, eds. Springer Verlag, New York. pp 1-15.

van Andel TH. (1994) *New Views on an Old Planet. A History of Global Change*. 2nd edition. Cambridge University Press, Cambridge.

Veblen TT, Hill RS, Read J, eds. (1996) *The Ecology and Biogeography of Nothofagus Forests*. Yale University Press, New Haven.

Villagrán CM. (1995) Quaternary history of the mediterranean vegetation of Chile. In *Ecology and Biogeography of Mediterranean Ecosystems in Chile, California, and Australia*. Arroyo MTK, Zedler PH, Fox MD, eds. Springer Verlag, New York. pp 3-20.

Wagner WL, Herbst DR, Sohmer SH. (1990) *Manual of the Flowering Plants of Hawai'i*, University of Hawaii Press, Honolulu.

Walter H. (1985) *Vegetation of the Earth*. Third edition. Springer-Verlag, Berlin.

Watt WS. (1940) *Climatological Atlas of Australia*. Bureau of Meteorology, Melbourne.

WCMC (World Conservation Monitoring Centre). (1992) *Global Biodiversity: Status of the Earth's Living Resources*. Chapman and Hall, London.

Weber CA. (1990) Chile. In *International Handbook of National Parks and Nature Reserves*. Allin, CW, ed. Greenwood Press, New York.

Westman WE. (1982) Coastal sage scrub succession. Gen Tech Rep PSW-58. Berkeley, CA: Pacific Southwest Forest and Range Experiment Station, Forest Service, U.S. Department of Agriculture. pp 91-99.

Westman WE. (1983) Plant community structure—spatial partitioning of resources. In: *Mediterranean-Type Ecostystems. The Role of Nutrients*. Kruger FJ, Mitchell DT, and Jarvis JUM, eds. Springer Verlag, Berlin. pp 417-445.

White ME. (1986) *The Greening of Gondwana*. Reed Books, Frenchs Forest, NSW, Australia.

White ME. (1994) *After the Greening, The Browning of Australia*. Kangaroo Press, Kenthurst, NSW, Australia.

Wilcox K. (1996) *Chile's Native Forests: A Conservation Legacy*. Ancient Forests International, Redway, CA.

WMO, UNESCO. (1970) *Climatic Atlas of Europe*. Technical supervisor, Steinhauser F. Cartographia, Budapest.

Worster D. (1988) The vulnerable earth: toward a planetary history. In *The Ends of the Earth. Perspectives on Modern Environmental History*. Worster D, ed. Cambridge University Press, Cambridge. pp 3-20.

Wrigley JW, Fagg M. (1989) *Banksias, Waratahs, and Grevilleas and All Other Plants in the Australian Proteaceae Family*. Collins, Sydney.

Zedler PH. (1986) Closed-cone conifers of the chaparral. Fremontia 14(3):14-17.

Zohary D, Hopf M. (1994) *Domestication of Plants in the Old World*, 2nd edition, Oxford University Press, Oxford.

INDEX

*Pages cited in **bold** include figures, explanations, or more detailed text.*

Abies concolor, 83

Acacia, 139, **147**, **148**, **151**, 156

   acinacea, **151**

   caven, 94, **96**, **97**, **98**, 101

   pentadenia, 165

acknowledgments, **vii**

acorns, **82**, **83**

Adelaide, 144, **225**

Adenostoma

   fasciculatum, **36**, 48, 49, 67, 68, 69, **70**, **71**

   sparsifolium, 69

Adesmia, 96, 102, 105

Aesculus californica, 79

afrikander, 138

Agapanthus, 138

Agathosma ciliata, 48

Agonis flexuosa, 165

agriculture, 42, 52, 78, 101, 147, **148**

Aguas Claras Ravine, 214

Albany, 143, 157, 160, **225**

Aldinga Scrub Conservation Park, 226

Aleppo pine, 139, 175, 182, **200**, **201**

alerce, **116**

Alerce Andino National Park, 216

allelopathy, **68**

Alstroemeria, 30, 94, **102**, 112

altitude, **95**, **195**

Amaryllidaceae, **138**, **188**

Amaryllis belladonna, 138

amaryllis family, 125, **136**, **138**, **188**

Anacardiaceae, 98

añañuca, 102, 104

Andalucia, **229**

Andes, **90**, **91**, **92**, **93**

Andrew Molera State Park, 210

Anigozanthos manglesii, **159**, **161**

annuals, **41-43**, 45

ant, **123**

Antarctic Kingdom, **20**

Antarctica, 26

Araceae, 138

Araucaria, **116**

   araucana, 94, 95, **113**, **116**

Arbutus

   menziesii, 79, 83, 88

   unedo, 48, 88, **174**, 177, **179**, **180**

Arctic Kingdom, **19**

Arctostaphylos glauca, 30, 48, 68, **69**, **72**, **73**, **75**

Arecaceae, 94, **108**, **112**, 172, **173**, **181**

Aristotelia chilensis, 96

aromatic plants, 185

arrayán, 115

Artemisia californica, 76

arum lily, 138

asphodel, **186**

Asphodeline lutea, **186**

Asphodelus aestivus, **186**

Asteraceae, 125, 135, **137**, **138**

Astroloma conostephoides, **159**

Athens, **12**, 13

Augusta, 225

Australia, 2, **143** (see also South Australia and Western Australia)

   dominant plant genera, **147**

   fire, **154**

   genera with abundant species, **148**

invasive plants, **162**

   loss of native vegetation to agriculture, **147**

   population density, 3, 4

   travel, **221**

   unusual pollinators, **154**

   variation in species over short distances, **156**

Australian heath family, **158**, **159**

Austrocedrus chilensis, 95, 112

Azara, 96

Babiana pulchra, 135, **136**, 162

Baccharis pilularis, 76

Bahia ambrosioides, 94, 105

bamboo, 94, 115

Banksia, 30, 122, 125, 148, **152**, 155, 163

   coccinea, **153**, 155

   flower structure, **152**

   grandis, 165

   marginata, **153**, 154

   ornata, 154

   seed cone, **154**

batha, 49, **185**

bay laurel, 79, 85, 88, 112

Belair National Park, 225

bell-fruited mallee, **150**

Bellis annua, 30

Betty's Bay, 220

Big Basin Redwoods State Park, 206

Big Sur, **50**

bigberry manzanita, 69, **72**, **73**

binomial nomenclature, **55**

bishop pine, **37**, 88

black oak, 55, **82**, **83**

black sage, 76

blue lechenaultia, **161**

blue oak, **51, 81, 82, 83**

boldo, 94, **112**

Bomarea salcilla, **105**

Books, travel and natural history

    Australia, **221**

    California, **204**

    Chile, **212**

    Mediterranean Basin, **227**

    Western Cape, **216**

    where to buy, **203**

Boronia edwardsii, **159**

Bot river protea, **127**

Botanic Gardens of Adelaide, **225**

bottlebrush, **150**

Brabejum stellatifolium, 48, **139**

Brodiaea, 30

Bromeliaceae, 106

bromeliads, 106

brown stringy bark, 166

bulbs, **41, 45, 135, 186**

bull banksia, 165

bush lupine, 76

bushwalking, **222**

Busselton, 199

cacti, **108, 109**

Calabria, Parco Nazionale della, 234

Caladenia,

    flava, **166**

    patersonii, **166**

Calaveras Big Tree State Park, 208

Calceolaria, 94, 102, **104**, 112

Calendon, 130

Calendon Gardens, 221

Calicotome villosa, 30, 185

California, 2, **59, 104**

    Central Valley, 59, 62, **64**

    chaparral, **63**

    climate, **59**

    Coast Ranges, **64**

    coastal scrub communities, **50**

    elevation, 61, 64

fire, **70**

fog, 3, 8, 61, 62, 93

forest, **84**

landscape, **59**

leaves, **48**

oak woodlands, **51**, 76, 78

plant list, **89**

population density, 3

rainfall, **60**

temperature, **61, 62**

topography, **58**

travel, **204**

vegetation, **63**

    mosaic, **66, 67**

California bay, 79, 85, 88, 110

California buckeye, 79

California holly, 74

California lilac, 30, 48, 67, **75**

California poppy, 74, **79**

California sagebrush, 76

calla lily, 138

Callistemon, 151

Calodendrum capense, 139

Calocedrus decurrens, 83

Calothamnus, 150

canelo, 94, 96, **114**, 115

canyon oak, **79, 82, 83**

Cape chestnut, 139

Cape Floral Kingdom, **21**, 119

Cape Leeuwin, 225

Cape of Good Hope, **218**

    Nature Reserve, **218**

Cape Province (see also Western Cape)

Cape strawflower, **137**

Cape Town, **12**, 13, 120, 135, 139, **218**

carob, 101, **181**

Castanea sativa, 179

Castle Rock State Park, **47**, 206

Casuarina, 154, **163**

    decussata, **165**

    pusilla, 154

    stricta, **165**

Ceanothus, 30, 67, **75**

chaparral, **68, 69**

    greggii, 48

    leucodermis, 48, 74

Cedarberg Mountains, 140, **219**

Central Chile (see also Chile), **91**

Ceratonia siliqua, **181**

Cercis occidentalis, 67, **75**

    siliquastrum, **180, 181**

chagual, 94, **106**

Chamaerops humilis, 172, **173**

Chamise, 67, 69, **70, 71**

    flammability, 70

    chaparral, 68, 74

    recovery after fire, 36

chaparral, 45, **47**, 63, **67**, 70, 97, 124, 177

    California lilac, **68**

    Ceanothus, **69**

    chamise, **68**

    manzanita, **72**, 74

    recovery after fire, **70, 72**

    scrub oak, **74**

    word origin, 67

chaparral-like shrublands, **45-47**, 177

checker bloom, **79**

chestnut, **179**

Chile, **91**

    Central Valley, **91**,93

    climate, **91**

    coastal matorral, **50**, 92, 96, **105**

    Coast Ranges, **91**, 93

    comparison with California, 97

    fire, **106**

    fog, 3, 93

    genera shared with California, **102**

    geography, **91**

    herbaceous species, **102**

    invasive Mediterranean grasses, **101**

    landscape, **91**

    leaves, **48**

    matorral, 92, **94, 96, 97**

north and south exposures, 96

plant list, 117

population density, 3

rainfall, 60, 92

ravine plants, 96

temperature, 61, 92

topography, 90

travel, 212

vegetation, 92, 93

vines, 102, 105

woodland and forest, 51, 94, 109

  sclerophyll, 109

Chilean fire bush, 94, 114, 115, 125

Chilean palm, 108, 112

Christmas candle, 138

chupalla, 107, 108

Chusquea, 94, 115

Cinque Terre, 233

Cistus, 30

  crispus, 183

  ladanifer, 183

  monspeliensis, 182

  salviifolius, 48, 174, 183

citrus family, 158, 159

Clanwilliam, 219

Clanwilliam cedar, 140

Clematis pubescens, 165

climate, 1

  charts, 12

  classification, 6

  coastal, 4

  continental, 4, 5

  coolest month, 11

  driest month, 9

  drought duration, 8, 12

  elevation, 5

  hottest month, 11

  maritime, 4, 5

  mediterranean, **xi**, 3, 8

    latitude, 4

    origins, 20, 21, 22, 24, 25

    types, 8

  microclimate, 62, 120, 121

mild, wet winters, 42

rainfall, 9, 12, 16, 60, 92, 120

season of peak rainfall, 12, 169, 170

temperature, 11, 61, 92, 121, 171, 173

variability, 13, 14

water deficit, 12

wettest month, 9

winter rain, 12

zones, 6, 7, 8

closed-cone conifers, 88, 198

coast live oak, 79, 81, 82, 83, 85

Coast Ranges, 51

coast redwood, 84, 85, 206

  root system, 85, 87

  stump sprouting, 87

  sun foliage, 84

coastal

  mallee, 45, 146, 159

  matorral, 45, 105

  sage scrub, 45, 76

    fire frequency, 76

  scrub, 45, 49, 50

  scrub communities, 50, 74

  vegetation, 50

coevolution, 132

coihue, 94, 114

coliguay, 48, 96, 101

Colliguaya odorifera, 48, 96, 101

Columbine Nature Reserve, 218

common asphodel, 186

common heath, 159

common smoke bush, 157

cone bush, 152, 157

Concepción, 92, 94

Conguillío National Park, 116, 216

conifers

  California, 88

  Chile, 116

  Mediterranean Basin, 191, 198

  Western Cape, 140

Conospermum, 47, 152

  stoechadis, 157

Conostylis, 159, 161

conservation, 52, 56, 57, 193

continental breakup, 25, 26

copihue, 105

coppice, 177, 178, 179

coral vine, 149, 165

cork oak, 172, 173, 194, 196

corm, 41

Correa, 158

  pulchella, 159

cowslip orchid, 166

coyote brush, 76

Cretaceous, 21

Cretan palm, 181

Crete, 10, 47, 51, 53, 194

  travel, 231

Crinodendron patagua, 114

Crocus, 188

  biflorus, 188

  sativus, 188

Cryptocarya alba, 48, 94, 96, 101, 110, 111

Crystal Cove State Park, 211

cup gum, 163

Cupressus macrocarpa, 87, 89

Cupressus sempervirens, 193, 200

Curtis's Botanical Magazine, 133, 188

cushion plants, 184, 185

Cuyamaca Rancho State Park, 211

cycad, 47

Cyclamen, 189

cypress, 148

Cyrtanthus ventricosus, 138

Cytinus, 183

daisy family, 125, 135, 137, 138, 160

Darwin, Charles, 26, 116

Darwinia, 151

De Hoop Nature Reserve, 221

deciduous, 28, 29, 33

deforestation, 192

Denmark, 225

dimorphic foliage, 33

Diuris, **166**

Dodecatheon, **81**

donkey orchid, **166**

Douglas-fir, 75, 78

Douglas iris, 74, **78, 79**

downy oak, **51,** 82, **197**

Drimys winteri, 94, 96, **114,** 115

Drosera, **156, 157**

drought, 9-12 (see also fire)
    Australia, 146
    California, **16,** 60
    duration, **8, 12, 16,** 43, 60
    Chile, 92
    Mediterranean Basin, 169
    stress, **13, 14,** 16
    evasion, **28, 29, 33**

drought-deciduous plants, **28, 29, 33**

Dryandra, 148, 152, 155
    nivea, **157**

dwarf fan palm, 172, **173**

elaiosome, 123, **154**

electron micrographs
    mycorrhiza, **39**
    sclerophyll and decidous leaves, **31**
    stomata, **31**

elevation
    California, 61, **64**
    Chile, **95**
    Mediterranean Basin, **195**

Embothrium coccineum, 94,**114, 115,** 125

endemic plants, **54, 56,** 156

Engelmann oak, **80, 82, 83**

Epacridaceae, 124, **158, 159**

Epacris impressa, **159**

Eremaea, 150

Erica, 124, 125, **130**
    adaptation to nutrient-poor soil, **133**
    arborea, 30, 48, **174, 177, 178**
    baccans, **133**
    cerinthoides, 133

coccinea, **132**
    flower structure, **131**
    geographic distribution, **130**
    grandiflora, **133**
    massonii, **132**
    patersonii, **132**
    pollination, **131, 132**
    propagation from seed, 133
    stomata, 131
    structure, **131**

Eschscholzia californica, 74, **79**

espinal, **97, 98, 101**

espino, 94, **97, 98, 101**

Eucalyptus, 30, **44,** 45, 49, 139, **147, 148,** 154, 156, **161, 162**
    baxteri, **166**
    calophylla, 163
    camaldulensis, **166**
    cnerifolia, 161, **164**
    cosmophylla, 161, **163**
    diversicolor, **53, 149,** 152, **164**
    flowers, **150**
    fruit, **163, 166**
    gomphocephala, 225
    incrassata, 161, **163**
    marginata, 155, **162, 164**
    obliqua, **166**
    oleosa, **161**
    preissiana, **150, 163**
    socialis, **161, 163**
    South Australia, 161, **163, 166**
    staeri, 155
    tetragona, 155
    wandoo, 163
    Western Australia, 161, **163**

Euphorbia, 125
    acanthothamnos, **173, 184, 185**
    dendroides, 185

everlastings, **42, 160**

evolution
    convergent, **46**
    of plants, **20**

family, **55**

fanflower family, **158**

farewell to spring, **74**

farming, 42, 52, 78, 101, 147, **148**

Fascicularia bicolor, **107**

featherflower, 151

Fernkloof Nature Reserve, 137, **220**

Festós, **232**

Figueroa Mountain, **180,** 211

fire, 11, **15,** 30, **35**
    Australia, **38, 154,** 182
    California, **36, 37, 70,** 76
    Chile, **106**
    in fynbos, **125**
    Mediterranean Basin, **182**
    natural vs. caused by humans, **70, 182**
    plant adaptation to, 29, **35, 36, 37, 38**
    plant recovery, **30, 36,** 69, 125
    renewal of vegetation, **36, 37,** 72
    soil insulation, **38**
    Western Cape, **125**

fire lily, 138

Fitzroy cypress, 95, **116**

Fitzroya cupressoides, 95, **116**

flame heath, **159**

flannel bush, 67, **75**

Flinders Chase National Park, **226**

Flinders Ranges National Park, 226

Floristic kingdoms, **18, 19**

food crops, **180**

forest, **45, 46,** 52
    Australia, **45, 146, 162**
    California, **84**
    Chile, **109**
    Mediterranean Basin, **176**
    Western Cape, 53, **139**

Foresta Umbra, **234**

France, 227

Fray Jorge National Park, 214

Freesia, 162

Fremontodendron californicum, 67, **75**

French lavender, 185

Fuchsia, 98

    lycioides, 30, 94, 102, 105

    magellanica, 94, **115**

fuchsia-flowered gooseberry, 74, **78**

fungus-root interaction, 39

fynbos, 45, **47**, 49, **50**, 121, 122, **124**, 125

fynbos/thicket mosaic, 122, **135**

Garden Route, **221**

Gargano, **234**

Garrapata State Park, 210

garrigue, 45, 49, 182, **184**, 186

Gastrolobium, 152

genera, species-rich, **148**

Genista hispanica, 177

genus, **55**

geophytes, **41**, 45, **135**, **186**

George, 140

Geraldton, 223

geranium, 135

giant sequoia, 87

Gladiolus, 30, 135, 138, 162

gold dust wattle, **151**

Gondwana, **25**, **26**, 115

Goodeniaceae, **159**

Goukamma Nature Reserve, 221

Granada, 230

granite outcrops, **157**, **160**

grape, 180

grass tree, **47**, **158**

grasses, invasive, 76, 110

grazing, 76, 98, **81**

Great Karoo, 123

Greece, 10, **47**, 51, 53, 194, **231**

Greek spiny spurge, **173**, **184**, **185**

Grevillea, 148, 152, 155

    annulifera, **157**

gum (see also *Eucalyptus*)

gum cistus, 183

Gunnera tinctoria, 94, **102**, **104**, **115**

Haemanthus coccinea, **136**

Haemodoraceae, **159**, **161**

Hakea, 139, 148, 154, 155

Hardenbergia, 165

    comptoniana, 152 129

Harold Porter Botanic Gardens, **53**, **220**

haustoria, **41**

heath, **130** (see also *Erica*)

Helichrysum, 42, 155

Helipterum, 30, **42**, 155

Henry W. Coe State Park, 206

herbaceous plants, **45**

Hermanus, 137, **220**

Heteromeles arbutifolia, 48, 69, 79

high maquis, **177**

Hippeastrum, 112

holm oak, 79, **174**, 177, **194**, 195

hoop petticoat narcissus, **188**

Hottentots Holland Mountains, 119

Hovea, 152

    elliptica, **149**, **165**

Humboldt Redwoods State Park, 206

Hunter-Liggett Military Reservation, 210

ice ages, **24**

Imbros Gorge, **233**

impact of man

    Australia, **162**

    California, **59**

    Chile, **101**

    Mediterranean Basin, xii, **192**

    Western Cape, **139**

insect-trapping plants, **40**, **156**, **157**

invasive plants

    Australia, **162**

    California, **78**

    Chile, 98, **101**

    Mediterranean Basin, **192**

    Western Cape, **139**

Iráklion, **232**

Iridaceae, 74, **78**, 125, **135**, **188**, **189**

Iris, **188**, **189**

    douglasiana, 74, **78**

    florentina, **188**

Iris family, 74, **78**, 125, **135**, **188**, **189**

ironwood, 140

island boronia, **159**

Isopogon, 152, **157**

Israel, 227

Italian cypress, 193, **200**

Italian man orchid, **190**

Italian Riviera, 233

Italy, **233**

Ixia, 135

jarrah, 155, **162**, **164**

Jerusalem sage, **33**, **185**

Jubaea chilensis, 94, **108**, **112**

Judas tree, **180**

Julia Pfeiffer-Burns State Park, 210

Juniperus oxycedrus, 48

Kageneckia

    angustifolia, 101

    oblonga, 48

Kalbarri, 223

Kalbarri National Park, 223

Kalflora Botanical Garden, 224

Kamieskroon, 137, 219

Kangaroo Island, 147, 151, **202**, **203**, **226**

kangaroo-paw, **158**

Karoo, **221**

Karoo Botanical Garden, **221**

karri, **53**, **149**, 152, **164**

karri oak, 165

karri wattle, 165

Kennedia coccinea, **149**, 152, **165**

Kermes oak, **47**, 175, 177, 182, **194**, **196**

Kermes scale insect, **196**

king protea, **127**, **128**

Kings Canyon National Park, 208

Kings Park Botanical Garden, **42**, **222**

Kirstenbosch Botanical Garden, 130, 139, **218**

Kleinmond Coastal Nature Reserve, 220

Kniphofia, **136**, 138
    uvaria, 138

knobcone pine, 88

Knossos, 232

Knysna, 140

Knysna National Lake Area, 221

Köppen's climate zones, **6**, **7**

Kunzea, 150

kwongan, 45, **47**, 49, 122, **146**, **152**

La Campana National Park, **52**, **97**, **108**, 112, 113, **213**

La Doñana, **229**

Lamiaceae, **185**

La Serena, 92, 94, **214**

Langeberg, 119

Lapageria rosea, 94, **105**

laurel, 88, 110, 112

Laurus nobilis, 88, 110, 112

Lavandula
    stoechas, **174**, **185**
    dentata, **185**

leaves
    dimorphic, **33**
    drought-deciduous, **33**
    hairy, 32, **33**
    sclerophyll, **20**, **29**, **30**, **31**, **48**, 49
    vertical, **32**, **33**

Lechenaultia, **159**, **161**
    biloba, **161**
    formosa, **161**

Leptospermum, **151**
    myrsinoides, 154
    sericum, 151

Leucadendron, 125, **130**
    argenteum, **130**, 139
    salignum, 48
    xanthoconus, **128**

Leucospermum, 122, 123, **129**

cordifolium, **129**, 130
reflexum, **129**, 130

lightning, **10**, **70**, 106, 120, 146

lignotubers, **36**, **38**

Liliaceae, **138**, **186**

lily family, **138**, **186**

lily of the Nile, 138

Limnanthes, 30
    douglasii, **79**

Linnaeus, 126

Lithocarpus densiflora, 88

Lithraea caustica, 30, 48, 94, 96, **97**, **98**, **99**, 106, 181

litre, 30, 48, 94, 96, **97**, **98**, **99**, 106, 181

Little Karoo, 123

Lobelia, 94, **114**, **115**

long-leaved protea, **127**, **128**

Loranthaceae, 41

Los Angeles area, 211

Los Padres National Forest, 209

Los Paraguas National Park, 216

low maquis, **183**

Ludlow Tuart Forest, **225**

Luma apiculata, 94, 115

lupine, 76, **81**

Lupinus, **81**
    arboreus, 76

macchia, 49

madrone, 79, 83, 88

mallee, **38**, 146, **147**, **159**, 163
    fire, 30, **38**
    heath, **160**

Mangle's kangaroo paw, **159**, **161**

manzanita, 68, **72**, **73**, **75**

manzanita chaparral, **68**, **72**

maps
    Australia
        rainfall, **144**
        temperature, **145**
        topography, **142**
        vegetation, **146**
    California
        rainfall, **60**

temperature, **61**
topography, **58**
vegetation, **63**
vegetation transects, **64**

Chile
    rainfall, **60**
    temperature, **61**
    topography, **91**
    vegetation, **92**, **94**

climate types, **6**

continental movement, **25**

floristic kingdoms, **18**

Mediterranean Basin,
    distribution of cultivated olives, **175**
    drought duration, **171**
    plant distributions, **173**, **174**
    rainfall, **170**
        peak seasons, **169**, **170**
        dry months, **171**
    temperature, **171**
    topography, **168**
    vegetation, **176**

mediterranean climate regions
    locations in world, **x**
    topography, **2**

Western Cape
    rainfall, **120**
    temperature, **120**
    topography, **118**
    vegetation, **121**

maquis, 45, 46, **47**, 97, **176**, **177**, 182, **191**

Maremma, **233**

Margaret River, 225

mariposita, 30

maritime pine, 139, 172, **173**, **198**, **199**

marri, 163

mastic tree, 181

matorral, **47**, 92, **96**, **97**
    effect of introduced rabbits, **98**

mayten, 94, **112**

Maytenus

boaria, 94, 112

oleoides, 140

meadow foam, 79

Mediterranean Basin, **xi**, 169

  agriculture, **192**

  conifers, **198**

  coppicing, **178, 179**

  drought duration, **171**

  farming, **192**

  fire, **182**

  forest, **176**

  garrigue, 50, **176**, 182, **191**

  geography, **168**

  impact of man, 192

  maquis, 45, 46, **47**, 97, **176**, **177**, **182**, **191**

  oaks, **176**, **191**, **194**

  origin of food crops, **180**

  plant species at meeting of continents, **172**

  plant distributions, **173**, **174**

  pollarding, **178, 179**

  population density, 3, 4

  rainfall, **168**

  temperature, **171**, **173**

  topography, **168**

  travel, **227**

  vegetation, **176**, **191**

  woodland and forest, **191**

mediterranean climate regions, **x**, **xi**, 1, 2

  areas, xii, 1, 2, 4

  latitudes, **x**, 2, 4, 176

  population, **4**

  population density, **4**

Mediterranean Sea, 1, **169**, 172, **173**, 176

Melaleuca, 150, 151, 173

  lanceolata, 165

messmate stringy bark, **51**, **166**

metric conversions, **258**

Metrosideros angustifolia, 48

microclimate

  California, **62**, **66**

  Chile, 96

Mimetes cucullatus, **128**, **130**

Mimulus, 74, **79**

mint family, **33**, **76**, **185**

mistletoe, 40

monkey puzzle tree, 94, 95, **113**, **116**

Monterey cypress, **87**, **89**, 209

Monterey pine, 3, **53**, **86**, **87**, **88**, **139**, 209

Monti del Gennargentu, 234

Mooney Grove Park, 208

Moore River National Park, 223

Morocco, 227

Mossel Bay, 120, 122

Mother Lode, 207, **208**

Mount Barker, **202**, **224**

Mount Etna, 234

Mount Lesueur, 156

Mount Lofty Range, 166

Mount Youktas, 232

Mount Tamalpais State Park, 206

mountain cypress, 95, 112

mountain rose, **128**, **130**

Muir Woods National Monument, 206

Murchison River Gorges, **223**

Museum of Western Australia, **223**

mycorrhiza, **39**, 139

Myrtaceae, 147, **148**, 177

myrtle, 147, **148**, 177

Myrtus communis, 177

Nahuelbuta National Park, 116, **215**

Namaqualand, 123, **137**, **219**

Namaqualand daisies, **137**

Narcissus, 30, **41**, 188

  bulbocodium, 188

  tazetta, 188

nasturtium, **105**

narrow leaf mallee, 161

narrow-leaved cistus, 182

Nerium oleander, **184**

New Almaden Quicksilver County Park, 206

nitrogen fixation, **40**, **69**

Norte Chico, 93

northern coastal scrub, **76**

Nothofagus, 45, 49, 95, **113**

  dombeyii, 94, **114**

  obliqua, 94, **110**, **114**

notro, 94, **115**

nutrient-deficient soil, **39, 56**, 68, **122**

Nuytsia floribunda, 40

oak, 45, 49 (see also *Quercus*, individual varieties)

  California, **76**

  Mediterranean Basin, **176**, **191**, **194**

oak woodland, 45, 49, **76**, **176**

ocean currents, 1

Ocotea bullata, 122, 140

Olea

  capensis, 140

  europaea, **174**, **175**

oleander, **184**

oleander-leaved protea, **124**

olive, **174**, **175**

  boundaries of cultivation, **175**

olivillo, 101

Ophrys, **189**

  tenthredinifera, **190**

orange pincushion protea, **129**

orchid family, 30, 157, 165, **166**, **189**, **190**

Orchidaceae, 30, 157, 165, **166**, **189**, **190**

orchids, 30, 157, 165, **166**, **189**, 190

Orchis boryi, **189**, **190**

  italica, **189**, **190**

  pauciflora, **189**

Ornduff, Robert, **ix**

Outeniqua Mountains, 119, 221

Paarl Mountain Nature Reserve, **47**, **219**

Palestine oak, **196**

palms, 94, **108**, **112**, 172, **173**

palo de yagua, 30, 94, 102, 105

paperbark, **150**, **151**, **163**

pangue, **102**, **104**, **115**

paragua, **116**

parasitic plants, **40**

patagua, 114

pea family, 147, **151** (also see *Aca-cia*)

Pelargonium, 135

pelú, **114**, **115**

Pemberton, **225**

peppermint, 165

Perth, **12**, 13, 143, 161, **222**

peumo, 43, 94, 96, 101, **110**, **111**

Peumus boldus, 94, 112

Pfeiffer Big Sur State Park, 210

Phaenocoma prolifera, **137**

Phlomis fruticosa, **33**, **185**

Phoenix theophrasti, **181**

photosynthesis, 30, 43

phrygana, 49, **176**, **184**

Phylyrea
    angustifolia, 48
    media, 48

Phycella, 102

Phytophthora cinnamomi, **164**

pincushion protea, **130**

pine, 45 (see also *Pinus*)

pine nuts, **198**

pink winter currant, 78

Pinus
    attenuata, 88
    halepensis, 139, 175, 182, **200**, **201**
    lambertiana, 83
    muricata, **37**, 88
    pinaster, 139, 172, **173**, **198**, 199
    pinea, **198**
    ponderosa, **83**
    radiata, 3, **53**, 86, **87**, 88, 139, **209**
    torreyana, **211**

Pistacia lentiscus, **181**

plane tree, 184, **194**

plant list
    Australia, **167**
    California, **89**
    Chile, **117**
    Mediterranean Basin, **201**
    Western Cape, 23, **122**

plant
    adaptations, 29
    annuals, 30
    binomial nomenclature, 55
    communities, **45**
        Australia, 23, **148**
        California, **62**, **63**, **64**
        Chile, 23, **102**
        Mediterranean Basin, 22, **172**
        Western Cape, **141**
    conservation, 52, 56
    dimorphic, **33**, **185**
    diversity, 52, **55**
    drought evasion, 28, 33, **185**
    drought-deciduous, **28**, 30, 33, **185**
    endemic, **54**, **56**, 156
    evolution, **20**
    exotic, **78**, **98**, **101**, **139**, **162**, **192**
    fossils, **20**
    geography, **19**
    herbaceous, **45**, 102
    insect-trapping, **40**, 156
    invasive, **78**, **98**, **101**, **139**, **162**, **192**
    mosaics, **63**, **66**, **67**
    origins, **20**, **21**, **24**, **25**
        Australia, 23, 148
        California, 22, 102
        Chile, 23, **102**
        Mediterranean Basin, 22, **172**
        northern hemisphere, 22, 26
        southern hemisphere, 23, 26
        Western Cape, 23, 122
    rare and endangered, 52, **54**, **55**

seasonally dimorphic, **33**, **185**

species
    abundance, **54**, **55**
    distribution, **174**
    diversity, **52**, **54**, **56,156**
    endemic, **54**, **56**
    rare and endangered, **54**, **55**
    variation over small distances, **156**

species-rich genera, **130**, **148**

stomata, **30**, **31**

survival traits, **28**, **29**

symbiosis with fungi, **39**

Platanus orientalis, 184, **194**

plate tectonics, **26**
    northern hemisphere, **26**
    southern hemisphere, **26**

Podocarpus, 52, 122
    latifolius, 140
    salignus, 94

Point Lobos State Park, **53**, **209**

Point Reyes National Seashore, **206**

pollard, **178**, **179**

pollinators, **155**

polyanthus narcissus, **188**

population, 3, **4**

porcupine grass, 161

Porongurup National Park, **157**, 160

Porongurup Range, **157**, **160**

Port Elizabeth, 122

Portofino, 234

Portugal, 227

precipitation (see also rainfall)

prickly plume grevillea, **157**

Preface, Robert Ornduff, **ix**

Prosopis, 101, 102

Protea, 30, 122, **125**
    arborea, 48
    compacta, 127
    cynaroides, 125, **127**, **128**
    eximia, **127**, **128**
    laurifolia, 140
    longifolia, **127**, **128**

neriifolia, 124

nitida, 30, 125, 140

repens, 127

season of bloom, 127

structure of flowers, 126

protea family, 40, 125, 147, 148, 152

Proteaceae, 40, 125, 147, 148, 152

proteoid roots, 40

Pseudotsuga menziesii, 75, 88

Psoralea obliqua, 48

Pteronia nitida, 30

Pueblos Blancos, 230

Puerto Montt, 216

purple-leaved sage, 76

purple sage, 33

Puya, 105, 106

   berteroniana, 106, 107

   chilensis, 94, 106, 107

pypies, 138

pyrenean oak, 195

Quaternary, 20, 21

Quercus, 49

   agrifolia, 49, 79, 81, 82, 83

   calliprinos, 196

   chrysolepis, 79, 82, 83

   coccifera, 30, 47, 175, 177, 182, 196, 197

   douglasii, 51, 81, 82, 83

   dumosa, 30, 48, 69, 74, 76, 77

   engelmannii, 80, 82, 83

   ilex, 48, 79, 174, 177, 194, 195

   kelloggii, 55, 82, 83

   lobata, 80, 82, 83

   macrolepis, 197, 198

   pubescens, 51, 82, 196

   pyrenaica, 195

   suber, 172, 173, 196

Quillaja saponaria, 48, 94, 96, 98, 100, 101, 106

quillay, 48, 94, 96, 98, 100, 101, 106

rainfall, 4, 5, 9

   Australia, 144

   California, 60-62

   Chile, 60, 92

   Mediterranean Basin, 168, 169, 170

   Western Cape, 120

Rancho Santa Ana Botanic Garden, 211

ray-flowered protea, 127, 128

red bottlebrush, 128

red buckwheat, 67

redbud, 75

red coral vine, 152, 149

red flowering currant, 74

red mallee, 163

red shank, 69

redwood, 45, 84-88, 207

Redwood National Park, 207

Restionaceae, 122, 134, 158

restios, 122, 134, 158

Rethymnon, 232

rhizomes, 41

Ribes

   sanguineum, 74, 78

   speciosum, 74, 78

ridge-fruited mallee, 163

Río Clarillo National Reserve, 104, 213

river red gum, 166

Riviera di Levante, 233

roble, 94, 110, 113

rock rose, 183

Ronald W. Caspers Wilderness Park, 211

Ronda, 230

root growth, 34, 35

rosemary, 182, 185

Rosmarinus officinalis, 182, 185

Royal Botanic Gardens at Kew, 196

Rutaceae, 158, 159

sage-leaved cistus, 174

Salcilla, 105

Salix, 102

Salmon correa, 159

Salvia

   apiana, 76

   leucophylla, 76

   mellifera, 30, 76

   triloba, 185

Samaria Gorge, 232

San Bernardino Mountains, 59

San Diego area, 211

San Francisco, 12, 13, 14, 63, 205

San Francisco area, 63, 205

San Gabriel Mountains, 59

sand-heath, 49

Santa Barbara, 51, 211

Santa Barbara Botanical Garden, 75, 211

Santa Rosa Plateau Ecological Reserve, 211

Santiago area, 92, 213

Sarcopoterium spinosum, 184, 185

Sardinia, 48, 234

Satureja gilliesii, 48

savanna, 6, 7, 101, 210, 211

sawfly orchid, 190

scarlet banksia, 153, 155

scarlet monkey flower, 74, 79

Schizanthus pinnatus, 30

sclerophyll vegetation, 20, 29, 30, 31, 33, 45, 46, 48, 49, 51, 52, 67, 87, 92, 96, 109, 146, 148, 176, 177, 191

Scott Creek Conservation Park, 226

scrub, 45, 46

scrub-heath, 152

scrub oak, 30, 48, 69, 76, 77

scrub oak chaparral, 74

seasonal dimorphism, 33, 185

seed

   ash-stimulated germination, 36

   dispersal by ants, 123, 154

   disturbance dependence, 36

   dormancy, 36

   heat resistance, 36

   heat stimulation, 36

   heat survival, 36

scant production, 122
slow maturation, 122
storage in canopy, 38, 124, 139
Sequoia National Park, 208
Sequoia sempervirens, 45, 84, 85, 86, 87, 207
Sequoiadendron giganteum, 87
Serapias parviflora, 190
Serpentine National Park, 224
Serranía de Ronda, 230
Seville, 230
Shark Bay, 143
sheoak, 154, 163, 165
shrublands, 46
shrubs, 45, 46
    dimorphic, 33
    drought-deciduous, 28, 33
Sicily, 234
Sidalcea malvaeflora, 79
Sierra de Cazorla, 230
Sierra Morena, 230
Sierra Nevada, 195
silver banksia, 153
silver tree, 130, 139
Skilpad Wildflower Reserve, 137, 219
small-flowered tongue orchid, 190
smokebush, 152
soapbark tree, 48, 94, 96, 98, 100
soft chaparral, 74
soil
    nutrient-deficient, 39, 56, 68, 147
Sophora, 94, 114, 115
South Africa, 119 (see also Western Cape)
South Australia, 2, 51, 144, 156, 158
    fire, 154
    landscape and climate, 144
    plant communities, 146
    travel, 225
southern beech, 45, 49, 94, 95, 110, 113, 114
Southwest Botanical Province, 143
Southwestern Australia, 143

Spain, 195, 229
Spanish broom, 177
Spanish lavender, 174, 185
Sparaxis elegans, 136
species
    abundance, 54, 55
    binomial nomenclature, 55
    distribution, 174
    diversity, 52, 54, 56, 156
    endemic, 54, 56
    rare and endangered, 54, 55
    variation over small distances, 156
spider orchid, 166
Spili, 233
spiny broom, 185
Stellenbosch, 219
stinkwood, 122, 140
Stirling Range, 145, 156, 160, 224
Stirling Range National Park, 224
Stoebe plumosa, 48
stomata, 31
stone pine, 198
strandveld, 45, 121, 135
strawberry tree, 88, 174, 177, 179, 180
strawflowers, 42, 160
Strybing Arboretum and Botanical Gardens, 75, 79
Stylidium, 157
succulents, 123
sugarbush, 69, 74, 127, 181
sugar pine, 83
sundew, 156, 157
sundew family, 156, 157
sunshine protea, 125, 128, 130
Swartberg, 119
symbiosis, plants with fungi, 39

tabaco del diablo, 114, 115
Table Mountain, 119, 218
tall cactus, 108, 109
tanbark oak, 88
taproot, 34
tea tree, 151, 154, 165

temperature, 8, 10, 11, 12
    Australia, 145, 146
    California, 61, 62
    Chile, 61, 92
    Mediterranean Basin, 171, 173
    Western Cape, 120, 121
Temuco, 215
Tertiary, 20, 21
tevo, 30, 48, 94, 96, 102, 103, 106
thorny burnet, 184, 185
three-leaved sage, 185
Tilden Park Botanic Garden, 206
timber, 179
time line, 21
tomillares, 49
Torndirrup National Park, 225
Torrey Pines State Beach, 211
toyon, 69, 74, 79
Transverse Ranges, 59
Travel, 203
    books (see also books)
    California, 204
        Central Coast, 209
        North Coast, 206
        San Francisco area, 205
        Sierra Foothills, 207
        Southern California, 211
    Chile, 212
        north of Santiago, 214
        Santiago area, 213
        south of Santiago, 215
    Mediterranean Basin, 227
        Crete, 231
        Greece, 231
        guided tours, 228
        Italy, 233
        Spain, 229
    South Australia, 225
        Adelaide area, 225
        Kangaroo Island, 226
    Western Australia, 221
        guided tours, 222
        north of Perth, 223

Perth area, **222**
south of Perth, **224**
Western Cape, 2, **216**
Cape Town area, **218**
Garden Route, **221**
guided tours, **217**
Karoo, **221**
Namaqualand, **219**
Overberg, **220**
population density, 3
West Coast, **218**
Winelands, **219**
tree heather, **174**, 177, **178**
tree hovea, **149**, **165**
tree spurge, 185
Trevoa trinervis, 30, 48, 94, 96, **102**, **103**, 106
Trichocereus, 94, 105
chilensis, 96, **108**, **109**
litoralis, 109
Triodia irritans, 161
Tropaeolum, 94, 112
tricolor, **105**
Tsitsikamma Coastal National Park, 221
Tsitsikamma Mountains, 221
tulip, **186**
Tulipa, **186**
doerfleri, **186**
saxatilis, **186**
Turkey, 227
turpentine, 165
Tuscany, **233**

Umbellularia californica, 79, 85, 88, 110
umbrella pine, **198**
University of California Botanical Garden, **205**
Ursinia cakilefolia, 30, **137**, **138**

Valdivia, 216
Valle de Elqui, **215**
Valle del Encanto, **214**
valley oak, **80**, **82**, **83**

valonia oak, **197**, **198**
Valparaiso, **12**, 13
van Riebeek, 139
vegetation, 45 (see also individual regions)
Australia, 23, **148**
California, **63**
Chile, **92**, **94**
Mediterranean Basin, 22, **172**
mosaics, **46**, 63
natural vs. actual, 46
Western Cape, 23, **121**
vegetation maps (see also maps)
Ventana Wilderness, 209
Verticordia, 151
vines, **102**, **105**, **149**, **152**, **165**
Vitis vinifera, 180

Walker Bay Nature Preserve, **220**
Warren National Park, **225**
Watsonia, 135
borbonica, 138
galpinii, **136**
wattle, **147**, **148**, **151**
West Cape Howe National Park, **225**
West Coast National Park, 137, **218**
Western Australia, 50, **143**, 156, 158
landscape and climate, **143**
plant communities, **146**
travel, **221**
Western Australian Christmas tree, 41
Western Cape, **119**, 152
fire, **125**
landscape and climate, **119**
nutrient-deficient soil, **122**
plant communities, **121**
travel, **216**
wildflower shows, 217, 218
western redbud, 67
Wheatlands, 224
white fir, 83

white-leaved sage, 76
Widdringtonia cedargergensis, 140
wild almond, 139
wild coastal fuchsia, 30, 94, 102, 105
Wilderness National Park, 221
wildfire (see also fire)
William Bay National Park, 221
winter's bark, 94, 96, **114**, 115
Wittunga Botanical Garden, 225
woodland, **44**, **49**, **51**, 52, 56
Australia, **146**
California, **76**, **84**
Chile, 51, 52, **109**
Mediterranean Basin, **191**
Western Cape, **139**
woodland management, 177, **178**, **179**

Xanthorrhoea, **47**, 154, **158**

yellow asphodel, **186**
yellowwood, 122, 144
Yosemite National Park, 80, **208**

Zantedeschia aethiopica, 138
Zapallar, 51

METRIC CONVERSIONS

TEMPERATURE

°Fahrenheit	°Centigrade
23	-5
32	0
41	5
50	10
59	15
68	20
77	25
86	30
95	35

°F = (°C x 1.8) + 32   °C =(°F - 32) x 0.56

DISTANCE

Miles	Kilometers
60	100
100	160
600	1000
1000	1600

miles x 1.61 = kilometers

kilometers x 0.62 = miles

1 degree of latitude at 37° = 69 miles or 111 kilometers

PRECIPITATION

Inches	Centimeters
4	10
10	25
20	50
40	100
100	250

inches x 2.54 = centimeters

centimeters x 0.39 = inches

HEIGHT

Feet	Meters
3.3	1
5	1.5
10	3
50	15
100	30

feet x 0.30 = meters

meters x 3.28 = feet

AREA

Acres	Hectares	Square Miles	Square Kilometers
1	0.4	400	1000
2.5	1	1000	2600

acres x 0.405 = hectares

hectares x 2.47 = acres

square miles x 2.59 = square kilometers

square kilometers x 0.38 = square miles

1 hectare = 100 meters x 100 meters

1 square kilometer = 1000 meters x 1000 meters or 100 hectares